Etienne Yergeau (Ed.)

Advanced Techniques for Studying Microorganisms in Extreme Environments

Life in Extreme Environments

Life in Extreme Environments

Series Editor
Dirk Wagner

Volume 8

Advanced Techniques for Studying Microorganisms in Extreme Environments

Edited by
Étienne Yergeau

DE GRUYTER

Editor
Etienne Yergeau
Université du Québec
Institut national de la recherche scientifique
Centre Armand-Frappier Santé Biotechnologie
531 boulevard des Prairies
Laval, Qc, H7V 1B7
Etienne.Yergeau@iaf.inrs.ca

Series Editor
Prof. Dr. Dirk Wagner
GFZ German Research Centre for Geosciences,
Helmholtz Centre Potsdam
Section Geomicrobiology
Telegrafenberg
14473 Potsdam, Germany
dirk.wagner@gfz-potsdam.de

ISBN 978-3-11-052464-2
e-ISBN (PDF) 978-3-11-052578-6
e-ISBN (EPUB) 978-3-11-052536-6
ISSN 2197-9227

Library of Congress Control Number 2019946858

Bibliographic information published by the Deutsche Nationalbibliothek
The Deutsche Nationalbibliothek lists this publication in the Deutsche Nationalbibliografie; detailed bibliographic data are available in the Internet at http://dnb.dnb.de.

Cover image: Luchschen/iStock/Thinkstock
Typesetting: Compuscript Ltd., Shannon, Ireland
Printing and binding: CPI books GmbH, Leck

www.degruyter.com

Preface

What is an extreme environment? For humans, this probably amounts to environments that would not allow them to live comfortably in, such as the bottom of the oceans, the south Pole, and the middle of extremely hot deserts. For a microbe, it is a different story. A temperate soil subjected to daily freeze-thaw cycles in the spring might be more extreme than permanently frozen soils in northern Canada. Otherwise, nonextreme environments that are subjected to constant human perturbations can become extreme for microbes (see Chapter 4 for a compelling case by Howard et al.).

But is the study of microbes occurring in extreme environments fundamentally different from the study of microbes occurring in more temperate settings? In most cases, probably not, and the techniques presented in this book can probably be applied to other environments as well. However, extreme environments do have some peculiarities. For one, they are generally less well studied than other types of microbial habitats, making the identification of genes and taxa more difficult when relying solely on molecular techniques. This has not stopped numerous researchers to perform such molecular-based studies in extreme environments as reviewed by Cowan and colleagues in Chapter 5. Along the same lines, in Chapter 6, Tremblay describes a modular analytical pipeline that can analyze sequencing data of any kind and can accept most databases for annotation. This approach was successfully used to characterize microbial communities in extreme environments, like ice, arctic soils, and extremely polluted soils, but was also shown to be useful for microbial communities in less extreme environments. Advanced isolation techniques, as comprehensively reviewed by Marcolefas and colleagues in Chapter 1, might help bridge the abovementioned gap by allowing more studies on isolated "extreme" microbes. Isolation and characterization are also prerequisites for the use of these microbes in industrial settings. Indeed, in Chapter 3, Schultz and Rosado show that extreme environments have been an almost inextinguishable source of microbes, genes, and protein that have desired characteristics for various industries.

Another key difference between temperate and extreme environments is that the latter environments are often so extreme that extracting molecules and microbes from the environmental matrix can become challenging. In Chapter 2, Choe and Lee expand on techniques to study microbes in rock and their habitats and show that even these microbes can be studied using modern techniques drawing from both the physical and biological sciences. One could also imagine that environments with extreme salt or pH values can offer challenges for the extraction of nucleic acids or other molecules using traditional methods.

Finally, in these times of global climatic changes, there is a need to understand the drivers of diversity, and I argue that extreme environments could be used as model

https://doi.org/10.1515/9783110525786-202

for that purpose. In Chapter 7, Moroenyane and Yergeau review available statistical techniques to study microbes in extreme environments, with a focus on understanding microbial diversity. I hope that this book will help students and established researchers to have an idea of the most advanced techniques available and of their peculiarities when applying them to study microbes in extreme environments.

Étienne Yergeau

Volumes published in the series

Volume 1
Jens Kallmeyer, Dirk Wagner (Eds.)
Micobial Life of the Deep Bisospehere
ISBN 978-3-11-030009-3

Volume 2
Corien Bakermans (Ed.)
Micobial Evolution under Extreme Conditions
ISBN 978-3-11-033506-4

Volume 3
Annette Summers Engel (Ed.)
Micobial Life of Cave Systems
ISBN 978-3-11-033499-9

Volume 4
Blaire Steven (Ed.)
The biology of Arid Soils
ISBN 978-3-11-041998-6

Volume 5
Jens Kallmeyer (Ed.)
Life at Vents and Seeps
ISBN 978-3-11-049475-4

Contents

Contributing authors

Evelien Adriaenssens
Gut Microbes & Health ISP
Quadram Institute Bioscience
Norwich Research Park, Norwich
United Kingdom NR4 7UQ
evelien.adriaenssens@quadram.ac.uk

Ianina Altshuler
McGill University
Ste-Anne-de-Bellevue
Quebec, Canada H9X 3V9
ianina.altshuler@mail.mcgill.ca

Terrence H. Bell
Penn State University
Department of Plant Pathology and
Environmental Microbiology
317 Buckhout Lab
University Park, PA 16802
thb15@psu.edu

Yong-Hoe Choe
Korean Polar Research Institute, 26
Songdomirae-ro, Yeonsu-gu
Incheon 21990, Korea
yhchoe@kopri.re.kr

Don Cowan
University of Pretoria
Hatfield, South Africa
don.cowan@up.ac.za

Pieter De Maayer
School of Molecular & Cell Biology
University of the Witwatersrand
Johannesburg, South Africa
Pieter.demaayer@wits.ac.za

Mia M. Howard
School of Integrative Plant Science
Plant Biology Section
Cornell University
Ithaca, NY, USA
mmh284@cornell.edu

Laura M. Kaminsky
Department of Plant Pathology and
Environmental Microbiology
The Pennsylvania State University
University Park, PA, USA
lmk52@psu.edu

Andre Kessler
Department of Ecology and Evolutionary Biology
Cornell University
Ithaca, NY, USA
ak357@cornell.edu

Yoo Kyung Lee
Korea Polar Research Institute
Incheon, South Korea
yklee@kopri.re.kr

Pedro Lebre
Centre for Microbial Ecology and Genomics
Department of Genetics
University of Pretoria
Pretoria, South Africa
Pedro.Lebre@up.ac.za

Thulani Makhalanyane
Centre for Microbial Ecology and Genomics
Department of Genetics
University of Pretoria
Pretoria, South Africa
Thulani.Makhalanyane@up.ac.za

Evangelos Marcolefas
McGill University
Quebec, Canada H9X 3V9
evangelos.marcolefas@mail.mcgill.ca

Itumeleng Moroenyane
Institut national de la recherche scientifique
Centre Armand-Frappier Santé Biotechnologie
Building 70, 531 Boulevard des Prairies
Laval, Québec, Quebec, Canada H7V 1B7
Itumeleng.Moroenyane@iaf.inrs.ca

Jean-Baptiste Ramond
Centre for Microbial Ecology and Genomics
Department of Genetics, University of Pretoria
Pretoria, South Africa
jbaptiste.ramond@gmail.com

Alexandre Rosado
Instituto de Microbiologia Paulo de Góes
Ilha Cidade Universitária-s/n bl I ss-Cidade Universitária
373 Avenida Carlos Chagas Filho
Rio de Janeiro, RJ, Brazil 21941-970
asrosado@micro.ufrj.br

Junia Schultz
Instituto de Microbiologia Paulo de Góes
Ilha Cidade Universitária-s/n bl I ss-Cidade Universitária
373 Avenida Carlos Chagas Filho
Rio de Janeiro, RJ, Brazil 21941-970
jschultz@micro.ufrj.br

Julien Tremblay
National Research Council Canada
Energy, Mining and Environment
6100, Royalmount avenue
Montreal, Quebec, Canada H4P2R2
julien.tremblay@nrc-cnrc.gc.ca

Marla Trindade
Institute for Microbial Biotechnology and Metagenomics
Department of Biotechnology
University of the Western Cape
Cape Town, South Africa
ituffin@uwc.ac.za

Angel Valverde
Centre for Microbial Ecology and Genomics
Department of Genetics
University of Pretoria
Pretoria, South Africa
avalverdeportal@gmail.com

Surendra Vikram
Centre for Microbial Ecology and Genomics
Department of Genetics
University of Pretoria
Pretoria, South Africa
surendravikram1@gmail.com

Lyle G. Whyte
McGill University
21, 111 Lakeshore Road
Ste-Anne-de-Bellevue, Quebec
Canada H9X 3V9
lyle.whyte@mcgill.ca

Etienne Yergeau
Université du Québec
Institut national de la recherche scientifique
Centre Armand-Frappier Santé Biotechnologie
Building 70, 531 boulevard des Prairies
Laval, Quebec, Canada H7V 1B7
Etienne.Yergeau@iaf.inrs.ca

Evangelos Marcolefas, Ianina Altshuler and Lyle G. Whyte

1 Advanced microbial cultivation methodologies and their applicability in cryoenvironments

1.1 Introduction

Despite the significant strides made in modern science, the "great plate anomaly," a fundamental limitation in microbiology first observed over 119 years ago, remains largely unresolved [1]. The term "great plate anomaly" was coined by Staley and Konopka in 1985 and refers to the discrepancy between total bacterial cells in a given environment compared to how many could be cultured on artificial media [2]. It has been estimated that a mere 0.1 to 1% of the total microbial species in the biosphere can be isolated using classical cultivation techniques [2]. This limitation in microbial cultivation arises from a number of contributing factors, such as the lack of specific nutrients, inappropriate incubation temperatures, unsuitable pH or osmotic conditions, incorrect oxygen levels, missing growth factors secreted by syntrophic organisms, or any combination thereof [3].

The development of 16S rRNA gene sequencing led to the realization that the portion of cultivable microorganisms does not accurately represent the community of the native environment [4]. This holds true both in terms of diversity and abundance of species [5]. Instead, microorganisms capable of growing on artificial media represent the phylotypes that are best adapted to laboratory conditions [6]. Although culture-independent molecular approaches have significantly improved our knowledge of microbial diversity and ecosystem composition, the need to culture microbes prevails. Cultivation is required to better understand microbial physiology and ecological roles and to establish biotechnological applications [5, 7, 8]. Several ground-breaking discoveries in microbiology have relied heavily on microbial cultivation. These include revolutionary biopharmaceutical discoveries (such as penicillin, streptomycin, and teixobactin), agricultural applications (such as the use of the insecticidal strains of *Bacillus thuringiensis*), and molecular breakthroughs (such as the discovery of clustered regularly interspaced short palindromic repeats in archaea) [9–17].

The importance of cultivation is especially evident for understudied, cold-adapted microorganisms living in extreme cold environments (cryoenvironments) [18, 19].The main motivators of studying cold-adapted microorganisms include understanding the cold temperature limits of life, establishing their significance in astrobiology, and harnessing their molecular resources for biotechnological applications [20–23]. All of these areas of research inherently rely on microbial cultivation and isolation. Over 75% of Earth consists of low-temperature environments (≤15°C). Accordingly, a considerable portion of terrestrial microbial biomass is produced at temperatures <5°C [24, 25]. The central biochemical constraint to life in

https://doi.org/10.1515/9783110525786-001

cryoenvironments is the lack of available liquid water [19]. Without liquid water, it is impossible to mediate biochemical reactions, to conduct molecular transport, and to provide necessary osmotic gradients [19]. Most available liquid water in cryoenvironments exists only due to freezing-point depression imparted by different solutes – typically NaCl [19]. Consequently, most cryophilic microorganisms are also halophilic since osmotolerance expands their survival in sub-zero temperatures [26, 27]. One such example includes *Planococcus halocryophilus* Or1, an aerobic Gram-positive coccoid bacterial strain that was initially isolated from high Arctic active-layer permafrost [27]. This isolate has been shown to grow at –15°C, remains metabolically active at –25°C, and can withstand high salinity exposure up to 19% NaCl [28]. Notably, this discovery would not have been possible without the isolation and cultivation of this strain [28].

Studying the cold limit of microbial life involves understanding how microbes living in cryoenvironments have evolved to cope with lower enzymatic activity, altered cellular transport, decreased membrane fluidity, and protein denaturation [29]. Cold-adapted microbes produce higher levels of unsaturated methyl-branched fatty acids and short acyl-chain fatty acids to maintain membrane fluidity [29]. Cold-shock proteins also help these microbes to maintain proper intracellular protein folding and inhibit ice crystal formation [29]. Additionally, at lower temperatures, water molecules become more ordered and do not associate with protein structures causing denaturation [29]. In response to these stresses, cold-adapted microbes have adapted biomolecular features through evolutionary selection that provide increased structural flexibility compared to mesophilic or thermophilic proteins.

The physicochemical conditions of terrestrial cryoenvironments also represent relevant starting points in the search for extraterrestrial life [28, 30]. It is hypothesized that if microbial life can thrive in extreme environments on Earth, then perhaps similar forms of life may be supported on analogous extra-planetary bodies such as Mars, Europa, and Enceladus [31, 32]. For example, the cold and arid polar desert of the McMurdo Dry Valleys in Antarctica has similar climatic conditions found on Mars [33]. At the Earth's other pole, the Canadian High Arctic features some of the planet's highest-fidelity Martian analogue sites [34]. In particular, Expedition Fjord on Axel Heiberg Island consists of a polar desert climate containing unique geographic sites. These include perennially liquid-state hypersaline springs and polygonal terrains comparable to ones found on Mars [34, 35]. Studying cold-adapted microorganisms in these habitats therefore has direct astrobiological implications [34].

In addition to studying cold adaptation and exploring astrobiological relevance, the molecular resources expressed by cold-adapted microorganisms could be exploited in biotechnological applications such as the enzyme and biopharmaceutical industries [29, 36]. Enzymes from cold-adapted microbes are capable of catalyzing chemical reactions under extreme conditions that are often similar to the conditions necessary for industrial processes [38]. Owing to their high activity at low temperatures and thermolabiltiy, these enzymes represent innovative developments for sustainable and environmentally friendly industrial bioprocesses [29]. Further, bioprospecting cold-adapted microbes for medically useful substances represents a recent promising approach in discovering novel natural chemotherapeutic agents including antibiotics [37]. Cold-adapted microorganisms are an untapped source of biomolecules with unique chemical structures tailored by evolutionary adaptation [39]. Recently discovered secondary metabolites with antimicrobial activities expressed by cold-adapted microbes include a new angucyclinone called frigocyclinone produced by *Streptomyces griseus*. This organism was isolated from Antarctic soil and inhibits Gram-positive bacteria. Another example is the antimycobacterial pigment violacein expressed by *Janthinobacterium* sp. Ant5-2 (J-PVP), initially isolated from land-locked freshwater lakes in East Antarctica [40, 41]. Identifying and isolating new biomolecules heavily depends on *in vivo* screening of cultivable strains [42]. Improved cultivation techniques will therefore assist in expanding the portion of cultivable isolates that could be phenotypically screened for exploitable biomolecules [42, 43].

Despite their biotechnological applicability and the myriad of basic science questions requiring further investigation, cold-adapted microorganisms remain to be an understudied and underexploited source of genetic and molecular resources [24, 25]. Logistical challenges surrounding accessing their remote geographical habitats, coupled with their fastidious growth requirements, have been considered the main barriers in studying microbes from cryoenvironments [42]. Traditional laboratory conditions cannot be applied to cultivate a large proportion of microorganisms originating from cryoenvironments [42]. For these reasons, the necessity to develop increasingly sophisticated cultivation methodologies is crucial to effectively study the microorganisms living in the expanding terrestrial (and, perhaps in the future, extraterrestrial) habitats currently being explored [42].

Efforts to address the issue of the uncultivability of environmental microorganisms have either centered around developing novel culturing methods that better mimic the environment or have attempted to culture microbes in their natural environments *in situ*. In this chapter, recent developments in microbial cultivation will be reviewed. Namely, techniques involving *in situ* cultivation using semipermeable hydrophilic membranes, micro-scale high-throughput (HT) approaches, and single-cell encapsulation. In each section of this chapter, a different *in situ* culturing method will be outlined as well as the different environmental sample types that they have been applied to. Finally, each technique's applicability in cryoenvironments will be assessed.

1.2 Diffusion chamber

The diffusion chamber technique was initially developed to culture microorganisms within their natural environment [44]. The rationale of this approach involves propagating microbial cells inside a gellan-filled chamber separated from the external environment by semipermeable membranes. The membranes provide growing cells access to growth components and signaling compounds while metabolic waste can diffuse out of the cultivation chamber [44]. The method's central component consists of a stainless-steel O-ring filled with agar or gellan gum containing diluted environmental cells [44]. This forms a central plug in the center of the O-ring (Fig. 1.1) [44]. The central agar plug is overlaid with semipermeable hydrophilic polycarbonate membranes (with 0.03 μm pores) (GE Osmonics Labstore, Minnetonka, MN) that are attached using silicon glue to the top and bottom faces of the O-ring, thus forming a chamber [44]. The chamber can then be placed in a natural or artificial substrate. Using this technique, a larger proportion of cells present in a given environment may be cultured compared to classical techniques [44].

In an initial proof-of-principal study, Kaeberlein et al. (2002) employed the diffusion chamber to demonstrate that previously uncultivable microorganisms can be isolated using this technique. Sediment collected from an intertidal sandy tidal flat was used to prepare their diffusion chamber (Tab. 1.1) [44]. To simulate the native natural habitat, sediment was placed in a laboratory aquarium with re-circulating natural seawater [44].This initial study led to the isolation of two previously undescribed bacterial strains: MSC1 and MSC2. MSC1 was deemed to be an undescribed species with 93% sequence similarity to *Lewinella persica* (pertaining to class Sphingobacteria, phylum Bacteriodetes) [44], whereas the closest relative of MSC2 was *Arcobacter nitrofigilis*. Notably, the authors demonstrated that their strains did not grow alone on conventional media without the presence of other syntrophic microorganisms on the same plate [44].

In a subsequent follow-up study, Bollman et al. (2007) compared the cultivable diversity between the diffusion chamber technique and classical artificial media. Near-shore sediment from a freshwater pond was placed in a continually aerated

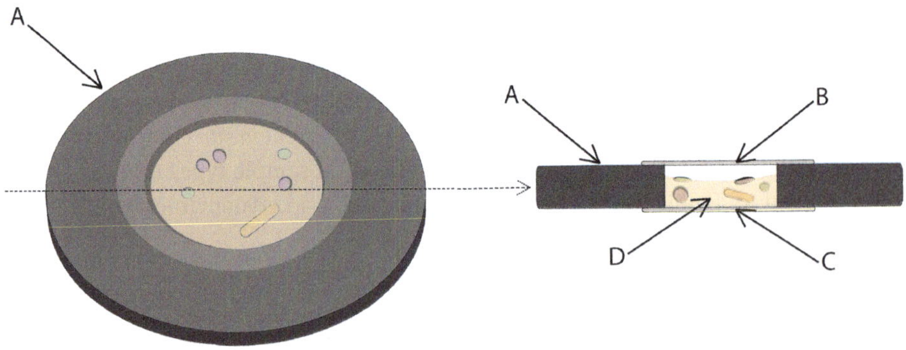

Fig. 1.1: Diffusion chamber. (A) Stainless-steel O-ring. (B–C) Semipermeable hydrophilic polycarbonate membranes (with 0.03 μm pores). (D) Propagating environmental microorganisms within central agar plug. Illustration adapted from Bollmann et al. (2007).

Tab. 1.1: Recent cultivation methodologies and key findings.

Cultivation method	Habitats studied	Cultured Microorganisms and notable isolates	Reference
Diffusion chamber	Sediment and freshwater samples from Turtle Pond, Boston, MA	Alpha-, Beta-, Gamma-, and Deltaproteobacteria; Bacteroidetes; Actinobacteria; Firmicutes; Spirochaetes; Acidobacteria; and Verrucomicrobia	Bollmann et al. (2007) [45]
	Subsurface sediments contaminated with uranium, heavy metals, nitrate, and organic pollutants from the Field Research Center (FRC) in Oak Ridge, TN	Alpha-, Beta-, Gamma-, Proteobacteria; Actinobacteria; Bacteroidetes; and Firmicutes	Bollmann et al. (2010) [46]
	Sediment from an intertidal sandy tidal flat (near the Marine Science Center, Northeastern University, Nahant, MA)	Strain MSC1 (93% sequence similarity to *Lewinella persica*) Strain MSC2 (closest relative, *Arcobacter nitrofigilis*)	Kaeberlein et al. (2002) [44]
Isolation chip	Seawater from the flow through seawater system at the Marine Science Center of Northeastern University, Nahant, MA (42°26 N, 70°56 W)	Firmicutes, Deltaproteobacteria, and Spirochaetes	Nichols et al. (2010) [51]
	Soil samples from a waterlogged wetland Marine Science Center of Northeastern University, Nahant, MA (42°26 N, 70°56 W)	Firmicutes and Gamma- and Deltaproteobacteria	
	Grassland soil, Maine, USA	A new species of Betaproteobacteria, *Eleftheria terrae*	Ling et al. (2015) [54]
	Canadian High Arctic permafrost, Axel Heiberg Island, Nunavut (79°26′N)	Putatively novel Pedobacter sp. (98% closest similarity in GenBank)	Goordial et al. (2017) [55]
Micro-Petri dish	Rhine River at Wageningen, the Netherlands	Alpha-, Beta-, Gammaproteobacteria	Ingham et al. (2007) [57]
Microbial trap	Garden soil from Verrill Farm (Massachusetts, USA), Pasadena (California, USA), and pine soil from Dover (Massachusetts, USA).	Actinomycetes, *Dactilosporangium, Catellatospora, Catenulispora, Lentzea*, and *Streptacidiphilus*	Gavrish et al. (2008) [50]
Hollow-fiber membrane chambers	Intertidal sediments from a tidal flat in Tokyo Bay, Chiba Prefecture, Japan (35°13 N, 139°79 E)	*Alphaproteobacteria, Gammaproteobacteria*	Aoi et al. (2009) [58]
	Activated sludge from the sewage wastewater treatment plant, Japan	Alpha-, beta-, and Gammaproteobacteria and Spirochaetes	
	Activated sludge from a laboratory scale enhanced biological phosphorus process, Japan	Alpha-, Beta-, and Gammaproteobacteria; Actinobacteria; and Bacteroidetes	

Tab. 1.1 (continued)

Cultivation method	**Habitats studied**	**Cultured Microorganisms and notable isolates**	**Reference**
Gel microdroplets	Mucus from *Fungia granulosa* corals, Gulf of Eilat, Inter-University Institute for Marine Science, Eilat, Israel	Alpha-, beta-, and gammaproteobacteria; *Actinobacteria*; *Bacteroidetes; Firmicutes; Gammaproteobacteria*; *Planctomycetes; Verrucomicrobia.*	Ben-Dov et al. 2009 [59]
	Soil from the Halutzah area of the Negev desert, Israel	Alpha- and gammaproteobacteria	
	Wastewater samples from a laboratory-scale wastewater bioreactor from the Ramat-Hovav industrial park in the Negev desert, Israel.	Betaproteobacteria	
	Tropical forest soil (05°56 N 00°03W) and (05°55N 00°03W), Ghana	Alpha-, Beta-, and Gammaproteobacteria and unspecified Gram-positive bacteria	Zengler et al. (2002) [60]
	Water samples, Sargasso Sea (31°50N 64°10W) and (32°05N 64°30W)	Alpha-, Beta-, and Gammaproteobacteria; Cytophaga-Flavobacterium-Bacteroides and relatives; and Planctomycetes	
In situ cultivation by tip	I-tips were embedded in *Lubomirskia baicalensis* sponges kept in aquariums at the Baikal Museum, Russian Academy of Science, Listvyanka, which were filled with water from Lake Baikal	Alpha-, Beta-, Gammaproteobacteria, Actinobacteria, Firmicutes	Jung et al. (2014) [61]
Soil substrate membrane system	Agricultural sandy loam soil Højbakkegaard and Sørningen, Denmark	Type II methanotrophs	Svenning et al. (2003) [62]
	Wetland soils Roskilde fjord, and from Skellingsted landfill, Western Sealand, Denmark	Type I and Type II methanotrophs	
	Garden soil, Macquarie University, Sydney, New South Wales, Australia	Beta- and Gammaproteobacteria, members of candidate division TM7	Ferrari et al. (2005) [64]

laboratory aquarium and covered with pond water (Tab. 1.1) [45]. This study led to the isolation of 112 species corresponding to Alpha-, Beta-, Gamma-, and Deltaproteobacteria; Bacteroidetes; Actinobacteria; Firmicutes; Spirochaetes; Acidobacteria; and Verrucomicrobia [45]. A high proportion of the isolated strains (26%) were deemed to be new species – where the majority (59%) of the "new" strains were isolated using the diffusion chamber technique [45]. In contrast, members from only Alpha-, Beta-, and Gammaproteobacteria; Bacteroidetes; Actinobacteria; and Firmicutes were cultured using standard plating [45]. Additionally, successive rounds of diffusion chamber incubations increased the overall recovery rate of isolates [45]. Certain strains initially cultured in diffusion chambers could be successfully isolated only on artificial media after successive rounds of diffusion chamber incubation. Although the mechanism behind this observation is unknown, successive diffusion chamber incubations "domesticate" microbial strains that would otherwise not grow on artificial media alone [45].

In a separate study, Bollman et al. (2010) evaluated whether the diffusion chamber approach was capable of isolating novel microorganisms tolerant to environmental stress factors present in polluted freshwater sediments (Tab. 1.1). Microbial recovery using the diffusion chamber method in subsurface acidic sediments contaminated with uranium, heavy metals, nitrate, and organic pollutants was compared to standard plating [46]. Using the diffusion chamber, 58 species were isolated corresponding to Alpha-, Beta-, and Gammaproteobacteria; Actinobacteria; Firmicutes; and Bacteroidetes [46]. Standard plating led to the cultivation of only Alpha- and Gammaproteobacteria, Actinobacteria, and Verrucomicrobia [46]. Unlike standard spread plating, the diffusion chamber approach led to the isolation of previously uncultured microbial isolates that exhibited tolerance to contaminated environmental conditions [46], namely, low pH and high concentrations of heavy metals and nitrate [46]. One notable isolate included OR221 (closely related to *Microbacterium laevaniformans*) that exhibited tolerance to all the stress factors present in the contaminated sediment at concentrations higher that those found in the original sediment sample [46].

1.3 Microbial trap

The microbial trap is an adaption of the diffusion chamber where semipermeable hydrophilic membranes with wider pores are used to trap filamentous microorganisms (such as actinomycetes and filamentous fungi) [47]. The device was developed by Gavrish et al. (2007) and consists of a sterile diffusion chamber (a washer containing a gellan gum center) covered by 0.2- to 0.6-µm pore-size polycarbonate membranes (47 mm diameter, Isopore, Millipore, USA) (Fig. 1.1). This uninoculated, sterile chamber is intended to be placed in a natural or engineered substrate. The expectation is that filamentous microbes will penetrate the sterile system through the wider membrane-pores using hyphal extensions. These cells subsequently become "entrapped" within the chamber and will propagate on gellan while allowed access to *in situ* nutrients and growth factors. Gellan gum has been shown to increase the rate of bacterial growth and supports growth of previously uncultured bacteria when compared to agar [48, 49]. Although the discrepancy in cultivability between agar and gellan is unknown, it is hypothesized that the sugars in gellan (glucose, glucuronic acid, and rhamnose residues) support the growth of a wider microbial diversity [47].

To assess whether the microbial trap technique was capable of selectively cultivating filamentous microorganisms *in situ*, Gavrish et al. (2007) compared the diversity of actinomycetes and filamentous fungi reared using the trap approach to conventional plating [50]. Additionally, this study explored the effect of using gellan gum instead of agar on microbial recovery [50]. Microbial traps were placed on top of garden and pine soil samples contained in a large petri dish and incubated in the dark for 14–21 days at room temperature (Tab. 1.1). The majority of the organisms in the trap were identified as actinomycetes, including rare groups such as *Dactilosporangium*, *Catellatospora*, *Catenulispora*, *Lentzea*, and *Streptacidiphilus*. These isolates were deemed to be unusual and infrequently cultured [50]. It was also observed that a higher variety of actinomycetes were cultured using the microbial trap approach as compared to Petri dish cultivation [50]. Lastly, the use of gellan gum instead of agar led to the recovery of a larger diversity of actinobacterial isolates [50]. This supports the body of evidence demonstrating that gellan gum sustains the growth of a larger variety of environmental isolates compared to agar [48, 49].

1.4 Isolation chip

The isolation chip (ichip) was initially developed in 2010 by Nichols et al. Building on the rationale of the diffusion chamber, this device offers an HT platform for parallel cultivation of environmental microbes – in essence consisting of 192 diffusion chambers in one single device [51]. This multichannel assembly allows for simultaneous isolation and culturing of environmental microbial strains [51]. It consists of three of hydrophobic plastic polyoxymethylene (DuPont, Delrin) plates [51]. The middle plate contains solid growth media where each well is inoculated with approximately 1 microbial cell from a diluted environmental sample (Fig. 1.2) [51]. The top and bottom layers are overlaid with 0.03-μm-pore-size, 47-mm polycarbonate membranes (Osmonics Inc., Westborough, MA), which cover the inoculated wells of the central plate [51]. Screws hold the three layers together and seal the assembly. Following inoculation, the ichip is placed in an environmental substrate for incubation. Only nutrients and growth factors from the native environment can penetrate and diffuse through the assembly, while propagating microbes cannot exit the assembly [51]. Since its initial description, the core design ichip has been adapted to better suit a variety of applications that require microbial cultivation. For instance, a miniaturized version of ichips made of surgical steel has been used to culture the human oral microbiota, whereas

NovoBiotic Pharmaceuticals uses a proprietary industrial-scale adaptation of the ichip as the principal cultivation tool to isolate and screen microorganisms for novel antimicrobials [52, 53].

In an initial study by Nichols et al. (2010), ichips were inoculated using seawater and soil samples where the culturable diversity was compared to parallel standard spread plating (Tab. 1.1). Among the seawater isolates, Firmicutes, Deltaproteobacteria, and Spirochaetes were the most frequently observed classes of strains cultured in the ichips, whereas Deltaproteobacteria, Firmicutes, and Gammaproteobacteria were the most frequently observed classes of strains cultivated from soil. In addition, it was demonstrated that 50% of the cells introduced to the channels managed to successfully form colonies. Species recovered from the ichip were observed to have significant phylogenetic novelty, 94 to 97% 16S rRNA gene sequence identity with known species [51]. Additionally, the culturable diversity recovered from the ichip approach was at least one order of magnitude greater than that achieved with standard in the Petri dish cultivation [51].

In a separate groundbreaking study, Ling et al. (2015) used the ichip method to discover teixobactin – a new class of antibiotics effective against drug-resistant

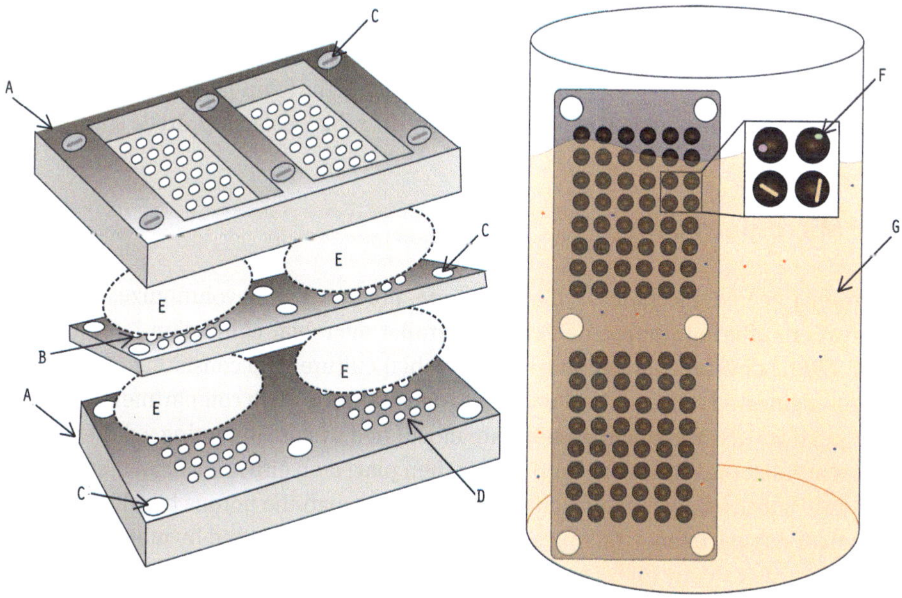

Fig. 1.2: Isolate chip (ichip). (A) Hydrophobic plastic polyoxymethylene plate (Du-Pont) (72 by 19 by 1 mm). (B) Central plate where environmental cells suspended in agar contained in through holes. (C) Screws and screw holes. (D) Through holes where nutrients can diffuse through the semipermeable membranes and reach propagating cells in the central plate. (E) Semipermeable hydrophilic polycarbonate membranes (with 0.03 μm pores). (F) Propagating environmental microorganisms. (G) Substrate (either environmental or simulated). Illustration adapted from Nichols et al. (2010).

Gram-positive bacteria. This unusual depsipeptide inhibits cell wall synthesis by binding to precursors of peptidoglycan and cell wall teichoic acid (lipid II and III, respectively) [54]. In this study, grassland soil was used to inoculate ichips (Tab. 1.1). Chemical extracts from 10,000 isolates reared in ichips were screened for antimicrobial activity against *Staphylococcus aureus*. The extract from a novel species of Betaproteobacteria, *Eleftheria terrae*, a class of bacteria not previously known to express antimicrobials, was found to express this new antibiotic molecule [54]. The discovery of this new antibiotic compound is a testament of how the discovery rate of new natural products will predictably increase as the capacity of novel culturing methods to isolate a larger portion microbial diversity broadens.

The first application of an ichip in a cryoenvironment was conducted by Goordial et al. (2017). In this study, a larger field prototype of the ichip (dubbed cryo-iplate) was used to cultivate microbes *in situ* within polygonal active layer permafrost soils (Tab. 1.1) [55]. This particular soil overlays ice wedges in polygonal permafrost analogous to terrains found on Mars. The cryo-iplate prototype design was inspired by the initial ichip by Nichols et al. (2010). Instead of using three hydrophobic plastic polyoxymethylene plates, this study used disassembled 200 µl-pipette-tip boxes to create the device assembly [55]. Using this technique, 39 microbial isolates were isolated including several that were identified as putatively novel through 16S rRNA gene sequencing [55]. Notably, a novel *Pedobacter* sp. (98% closest similarity in GenBank) was isolated [55]. This study demonstrates the versatility of the ichip-like devices in cryoenvironments and the potential of recovering new microbial isolates in Arctic soils.

1.5 Micro-petri dish

Micro-engineered mechanical platforms are predicted to revolutionize microbial cultivation, offering unprecedented HT, parallel microbial cultivation [56]. Ingham et al. (2007) developed a miniaturized microbial culture chip consisting of a porous micro-engineered ceramic sheet with up to one million growth compartments that are 7 × 7 µm (Fig. 1.3). Microbial dilutions are inoculated by spread-plating a diluted environmental sample. The micro-Petri dish is then placed on either artificial media or on a natural substrate, where nutrients can diffuse through the porous basal layer up to the micro-compartments. Colonies are extracted and subcultured by manual picking under the microscopy [57]. This technique allows HT microbial culturing and could be amenable to full automation. It could also be used to screen, count, and select colonies [57]. The micro-petri dish ultimately addresses the conventional Petri dish's incompatibility with automation and subdivides microbial growth compartments to the level of genuine HT [57].

Ingham et al. (2007) used the micro-petri dish to culture microorganisms from river water and to screen growing microcolonies for organic phosphate metabolism. Their inoculated micro-petri dish was placed on plates of filter-sterilized river water

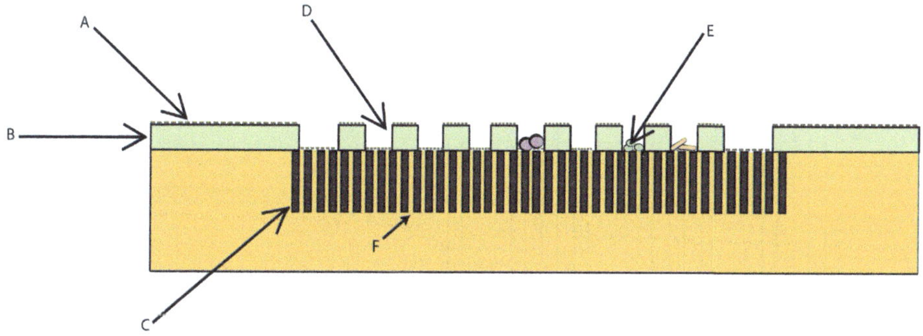

Fig. 1.3: Micro-petri dish. (A) Platinum layer. (B) Laminate layer. (C) Porous aluminum that allows diffusion of nutrients to compartments. (D) Compartment where cells propagate. (E) Cultured microbial cells. (F) Substrate (either environmental or simulated). Illustration adapted from Ingham et al. (2007).

solidified with 0.3% (w/v) Gelrite (Merck) to simulate the source environment (Tab. 1.1). This led to the growth of 207,000 microcolonies. To screen these microcolonies for organic phosphate metabolism, the contents of the micro-petri were transferred to slides with flurogenic substrate fluorescein 3,6-diphosphate). Microcolonies observed to convert the substrate to fluorescein were selected for 16S rRNA gene sequencing. This led to the isolation of *Aeromonas*, *Acidovorax*, *Klebsiella*, and *Rhodobacter* spp. Six previously uncultured organisms capable of phosphate metabolism were isolated. One pertained to a putatively uncultured genus, whereas the other had nearest relatives to *Puucibacter toxinivorans*, along with members of the genera *Dechloromonas*, *Sphaerotilus*, *Aeromonas*, and *Sphingomonas* [57].

1.6 Hollow-fiber membrane chamber

The hollow-fiber membrane chamber (HFMC) was designed for *in situ* incubation of microorganisms in aquatic environments [58]. The device is comprised of a 48- to 96-chamber unit system where each chamber unit consists of a porous hollow-fiber polyvinylidene fluoride (PVDF) membrane (Asahi Kasei Chemicals Co., Tokyo, Japan) (Fig. 1.4) [58]. Serially diluted environmental microbial cells are injected in the ports of each chamber while the sampling port is used to withdraw cultivated biomass at the end of the incubation period [58]. During incubation, the chamber system is immersed in either a natural or engineered aquatic environment [58]. The porous PVDF membranes allow the diffusion of nutrients, growth components, and signaling compounds while restricting movement of microbial cells inside or outside the system. This leads to dynamic cultivation conditions inside the device as they change according to external conditions [58]. The device can allow for the cultivation

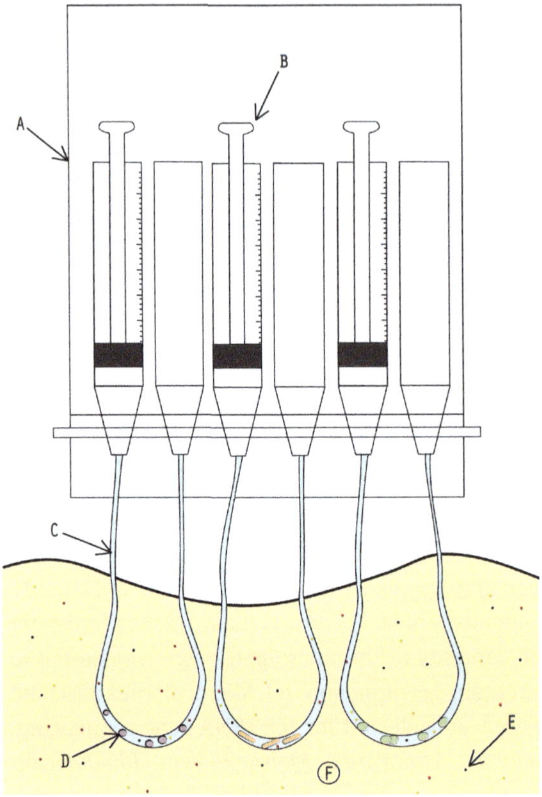

Fig. 1.4: Hollow-fiber membrane chamber (HFMC). (A) Cover encasing the 48–96-chamber unit system. (B) Syringe. (C) Hollow-fiber membrane. (D) Propagating cells. (E) Nutrients, growth components, and signaling compounds. (F) Substrate (either environmental or simulated). Illustration adapted from Aoi et al. (2009).

of fastidious microorganisms requiring syntrophic partners and specific low-nutrient concentrations from their environment [58].

A study by Aoi et al. (2009) was conducted to assess the diversity cultured by the HFMC technique as opposed to conventional agar plating. Several aquatic samples were used in this study: intertidal sediments sampled from a tidal flat, activated sludge from the sewage wastewater treatment plant (SWTP), and activated sludge from a laboratory-scale enhanced biological phosphorus process (EBPR) (Tab. 1.1). From the intertidal sediments, a total of five isolates corresponded to new phylotypes whereas four isolates were deemed to be new phylotypes from sewage activated sludge. From the laboratory activated sludge, 10 isolates were deemed to have been previously uncultured [58]. The proportion of cultivable cells recovered by the HFMCs were 12.3%, 21.0%, and 9.2% from the tidal flat, SWTP, and EBPR, respectively, whereas those from Petri dishes were 2.1%, 1.4%, and 0.2%, respectively [58].

1.7 Gel microdroplets

The gel microdroplet (GMD) approach involves encapsulating environmental microorganisms in polysulfone-coated agar spheres. The GMDs could then be incubated *in situ* or in simulated natural habitats. A main limitation with this technique involves the lack of a physical barrier between the GMD spheres and the environment. To address this, GMDs could be placed in growth columns or in mesh bags [59, 60]. Environmental samples are diluted such that approximately one bacterial cell is encapsulated in a polysulfone coated agar sphere. Cells within GMDs have access to naturally occurring nutrients and growth factors in their environment. After incubation, GMDs harboring microbial growth can be separated and retrieved using a flow cytometric cell sorter [59, 60].

In a study conducted by Ben-Dov et al. (2009), the GMD approach was used to culture microorganisms from various habitats and compared the culturable diversity to parallel conventional plating. GMDs were prepared using mucus from *Fungia granulosa* corals, desert soil, and wastewater samples obtained from a laboratory-scale wastewater (Tab. 1.1) [59, 60]. Polysulfone-coated agar spheres containing environmental cells were either placed freely or contained in mesh bags during a 3-week incubation period in source habitats [59, 60]. GMDs containing mucus associated microbes were incubated in aquaria containing the coral host species (*F. granulosa*), along with artificial seawater. GMDs prepared with wastewater-derived microbial cells were incubated in wastewater bioreactors, whereas GMDs prepared from soil-derived cells were placed in disposable cups filled with soil from the source site. To enrich and isolate strains, consecutive rounds of GMD cultivation were performed by aseptically transecting spheres containing growth with a sterile scalpel, melting them, and mixing with fresh warm agar to form new spheres. Overall, Ben-Dov et al. (2009) demonstrated that culturable diversity using GMDs is 10–100-fold greater than standard plating and 50% of isolated strains shared similarity between 85 and 96% with previously identified microorganisms [59, 60].

In a separate study by Zengler et al. (2002), GMDs were prepared using soil samples collected from tropical forest in Ghana and water samples from the Sargasso Sea (Tab. 1.1) [60]. During incubation, GMDs were loaded into sterile chromatography columns XK-16 (Amersham Pharmacia) containing simulated habitat substrate [60]. GMDs were incubated in the columns for a period of at least 5 weeks before being sorted with flow cytometry and subcultured into a 96-well microplate [60]. The cultivation techniques described above led to the isolation of Alpha-, Beta-, and Gammaproteobacteria, Gram-positive bacterial strains from tropical forest soil. Alpha-, Beta-, Gamma-, Deltaproteobacteria, as well as of the Cytophaga-Flavobacterium-Bacteroides relatives and Planctomycetes from Sargasso Sea water samples with previously uncultivated phylotypes within the Planctomycetales (Tab. 1.1) [60]. This study demonstrated that the GMD approach could recover less abundant members from both soil and marine environments.

1.8 In situ cultivation tip (I-tip)

The *in situ* cultivation by tip (I-tip) device was developed by Jung et al. (2014) in order to efficiently isolate microorganisms associated with aquatic sponges. Sponges host large communities of microorganisms, where most are host specific and largely remain uncultivated [61]. The device consists of a standard micropipette tip (Eppendorf, Germany) filled with microbeads (Sigma-Aldrich) and agar. The narrow end of the tip is then embedded in aquatic sponges (Fig. 1.5). Microorganisms associated with the invertebrate host invade the basal layer of microbeads at the narrow end of the tip and proliferate in the upper agar layer using nutrients from their native environment [61].

In an initial study introducing this cultivation method, Jung et al. (2014) demonstrated the efficacy in microbial recovery using the I-tip device. I-tips were embedded in *Baicalospongia* sp. and *Lubomirskia baicalensis* sponges and then incubated for 4 weeks before being aseptically cut open [61]. Subculturing was performed on Reasoner's 2A (R2A) plates using the I-tip agar layers. This resulted in the isolation of 34 species pertaining to Actinobacteria; Alpha-, Beta-, Gammaproteobacteria; and Firmicutes. Contrastingly, standard plate cultivation of the same sponges led to the cultivation of Betaproteobacteria, Firmicutes, and Gammaproteobacteria (Tab. 1.1). A chi-square test was used to compare the proportion of novel strains (<97% 16S rRNA gene similarity to closest known relative) obtained by the I-tip method as opposed to standard cultivation. This experiment demonstrated that it is more likely to culture more microbial novelty using the I-tip as opposed to standard culturing techniques [61].

1.9 Soil substrate membrane system

The soil substrate membrane system (SSMS) technique was developed to mimic the natural environment of soil and to select for previously uncultivated soil microorganisms. The SSMS techniques involves spreading soil dilutions directly on cell culture insert (Nunc A/S, Roskilde, Denmark) PCM membranes (Millipore, North Ryde, New South Wales, Australia) [62–64]. Underneath the cell culture insert membrane, wet, unsterilized soil is used as the only source of nutrients, allowing for oligotrophic microcolony formation. The PCM membrane provides support for microbial growth, acts as a barrier between the soil substrate and the growing colonies, and allows nutrients to diffuse from the substrate to the propagating cells (Fig. 1.6). Following incubation, microcolonies formed on the PCM could either be characterized by fluorescence *in situ* hybridization, used for DNA extraction, or subcultured [62–64]. To attain pure isolates, successive culturing with SSMS could be applied. This could be achieved by quartering a PCM with microbial growth (under sterile conditions with a scalpel) and suspending cells in 0.9% NaCl. The SSMS method can be applied to different soil

Fig. 1.5: *In situ* cultivation by tip (I-tip). A) Pipette tip (20–200 μl). (B) Waterproof adhesive separating tip contents and exterior environment. (C) Propagating cells. (D) Agar medium. (E) 50~100 or 150~212 glass beads. (F) Bottom aperture where microbial cells enter system. (G) Sponge host. Illustration adapted from Jung et al. (2014).

types and could be incubated at various time durations and temperatures tailored to optimize microbial recovery from different terrestrial environments [62–64].

Svenning et al. (2003) used the SSMS to isolate methane oxidizing bacteria from soil [63]. Agricultural sandy loam soil and wetland soils were used to prepare the SSMS (Tab. 1.1) [62]. To attain pure isolates, microcolonies from the first round of culturing were subcultured in a second round of SSMS by picking colonies using a 2 μl capillary tube and inoculating onto a new PCM [62]. Using this cultivation technique, type II methanotrophs were cultured from landfill soil while both type I and type II methanotrophs were cultured in the wetland soil [62]. Several isolates were found to carry the soluble methane monooxygenase gene. Additionally, 23 out of the 24 colonies that appeared after SSMS incubation were found to be different strains, highlighting the technique's efficiency in recovering a wide microbial diversity [62].

In a study by Ferrari et al. (2005), the SSMS approach was used to culture microcolonies pertaining to uncharacterized microorganisms. In this study, a larger portion of microbial diversity was hypothesized to be recovered by isolating colonies that require microscopic detection. This strategy was used since a considerable proportion of culturable microorganisms do not form macroscopic colonies [64]. To prepare the SSMS,

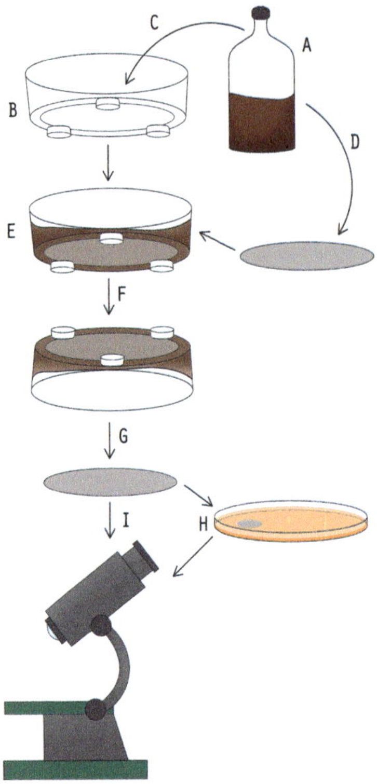

Fig. 1.6: Soil substrate membrane system (SSMS). (A) Diluted soil sample. (B) Cell culture insert with PCM. (C) Diluted soil slurry used as substrate in cell culture insert. (D) Inoculation. (E) PCM placed underneath soil slurry substrate. (F) Incubation of cells on PCM. (G) Removal of PCM after incubation. (H) Subcultivation on plate. (I) Observation of colonies on PCM with microscopy. Illustration adapted from Ferrari et al. (2008).

Australian garden soil was passed through a 2-mm-mesh sieve (Tab. 1.1). Subculturing was achieved by first observing the cultivated colonies under fluorescence microscopy, physically cutting each marked microcolony-membrane complex, and suspending the cells in a saline solution before transferring the cells to a secondary SSMS. Using this technique, *Methylobacilli* and *Pseudomonas* strains (with 96 to 100% similarity to closest relatives) and bacterial strains without cultivated representatives (including members of candidate division TM7) were cultivated [64]. The use of the SSMS technique in this study affirmed that a broader portion of microbial diversity may be isolated by focusing cultivation efforts on microorganisms that form microcolonies [64].

1.10 Applicability in cryoenvironments

Most of the recently developed cultivation techniques reviewed in this chapter were developed and applied under non-extreme conditions. Although not empirically tested herein, the applicability of the methods discussed in this chapter to cryoenvironments could be predicted by considering their material constituents' stability under the physicochemical conditions found in such environments. For any of the techniques described above to be suitable for *in situ* incubation in extreme cryoenvironments, these materials would have to withstand exposure to low temperatures, a range of pH, and high levels of salt (since most liquid water in cryoenvironments is saline) [18].

The diffusion chamber, the ichip, and the microbial trap all employ a similar cultivation rationale: microbial cells propagating on gellan gum separated from the exterior environment by hydrophilic polycarbonate membranes. Theoretically, these devices may be submerged in either liquid, semisolid, and solid substrates either *in situ* or in engineered environments. Given gellan gum's durable properties, these assemblies could be incubated at cold temperatures, alkaline or alkali (stable between pH of 3.5 to 8), and saline substrates (assuming gellan has been amended with salt while setting) without compromising the device's integrity [65, 66]. One limitation with these methods is their requirement for the diffusion of nutrients across their hydrophilic membranes, which requires liquid water [52]. Desiccating or complete freezing of the gellan gum would hinder these techniques in subzero arid environments. Goordial et al. (2017) showcased the versatility of an ichip-like device in cryoenvironments and represents the first attempt in applying such a method in an Arctic environment [55].

The micro-petri dish and the SSMS could also theoretically withstand low temperatures, a wide range of pH, and saline conditions given their ceramic and polycarbonate constituents [67–69]. However, both techniques could only be applied in solid or semisolid substrates since they both rely on nutrient diffusion from the basal surface passing through a porous membrane to reach the apical surface where microbial cells propagate. This would be incompatible with full submersion in a liquid substrate. Nevertheless, it would be possible to place these devices in a shallow liquid environment where they would rest on top of the liquid substrate, for example, on top of intertidal sediment where water would be in contact with the basal surface but not with the apical surface where cells are growing. The limitation of desiccation is once again a hurdle for both of these methods given that they rely on capillary diffusion of nutrients from the basal substrate to the cells growing on the apical surface (separated by a porous membrane) [52].

The HFMC, the I-tip, and the GMD techniques are inherently less versatile in the conditions they could be used in. The HFMC and the I-tip design could be applied only in aquatic substrates, while the I-tip could almost exclusively be used to culture microorganisms associated with aquatic invertebrate hosts. Nonetheless, these methods could theoretically be applied in aquatic cryoenvironments as they can foreseeably withstand cold temperatures, high salt concentrations, and to a lesser extent, a variety of pH conditions. Although GMDs have been used in a variety of habitats (such as marine and desert soils as described above), the use of agar in both the I-tip and in GMDs could predictably not withstand saline the and varying pH environments that are oftentimes associated with cryoenvironments [70, 71].

1.11 Concluding remarks

The recent culturing techniques explored in this chapter highlight the growing appreciation for advancing microbial cultivation methods. The central focus of these methods involves preserving the physical and chemical composition of the native environment during incubation. All methods efficiently recovered a wider breadth of microbial diversity when compared to classical plate cultivation. Putatively novel phylotypes were consistently discovered throughout the habitats discussed above. Although further empirical confirmation is required, the diffusion chamber, the ichip and the microbial trap are hypothesized to be the most versatile in terms of their applicability to cryoenvironments. Given the understudied status of cryoenvironments, the increasing use of these techniques in such habitats will predictably increase the discovery rate of previously uncultured microorganisms.

References

[1] Winterberg H. Zur methodik der bakterienzählung. Zeitschrift für Hygiene und Infektionskrankheiten. 1898;29(1):75–93.

[2] Staley JT, Konopka A. Measurement of *in situ* activities of nonphotosynthetic microorganisms in aquatic and terrestrial habitats. Ann Rev Microbiol. 1985;39(1):321–46.

[3] Vartoukian SR, Palmer RM, Wade WG. Strategies for culture of 'unculturable' bacteria. FEMS Microbiol Lett. 2010;309(1):1–7.

[4] Olsen GJ, Lane DJ, Giovannoni SJ, Pace NR, Stahl DA. Microbial ecology and evolution: a ribosomal RNA approach. Annual reviews in microbiology 1986;40:337–65.

[5] Dojka MA, Harris JK, Pace NR. Expanding the known diversity and environmental distribution of an uncultured phylogenetic division of bacteria. Appl Environ Microbiol. 2000;66(4):1617–21.

[6] Stewart EJ. Growing unculturable bacteria. J Bacteriol. 2012;194(16):4151–60.

[7] Ward DM, Weller R, Bateson MM. 16S rRNA sequences reveal numerous uncultured microorganisms in a natural community. Nature. 1990;345(6270):63–5.

[8] Hugenholtz P, Goebel BM, Pace NR. Impact of culture-independent studies on the emerging phylogenetic view of bacterial diversity. J Bacteriol. 1998;180(18):4765–74.

[9] Piddock LJ. Teixobactin, the first of a new class of antibiotics discovered by iChip technology? J Antimicrob Chem. 2015;70(10):2679–80.

[10] Höfte H, Whiteley H. Insecticidal crystal proteins of *Bacillus thuringiensis*. Microbiol Rev. 1989;53(2):242–55.

[11] Hoekema A, Hirsch P, Hooykaas P, Schilperoort R. A binary plant vector strategy based on separation of vir-and T-region of the Agrobacterium tumefaciens Ti-plasmid. Nature 1983;303:179–80.

[12] Karginov FV, Hannon GJ. The CRISPR system: small RNA-guided defense in bacteria and archaea. Mol Cell. 2010;37(1):7–19.

[13] Ishino Y, Shinagawa H, Makino K, Amemura M, Nakata A. Nucleotide sequence of the iap gene, responsible for alkaline phosphatase isozyme conversion in Escherichia coli, and identification of the gene product. Journal of bacteriology 1987;169:5429–33.

[14] Fleming A. On the antibacterial action of cultures of a penicillium, with special reference to their use in the isolation of *B. influenzae*. Br J Exp Pathol. 1929;10(3):226–36.

[15] Waksman SA. Streptomycin: background, isolation, properties, and utilization. Science. 1953;118(3062):259–66.

[16] Jones W, Leigh J, Mayer F, Woese C, Wolfe R. Methanococcus jannaschii sp. nov., an extremely thermophilic methanogen from a submarine hydrothermal vent. Archives of Microbiology 1983;136:254–61.

[17] Kpn I, Sma I, Sst I. Complete genome sequence of the methanogenic archaeon, *Methanococcus jannaschii.* Science. 1996;273:23.

[18] Goordial J., Lamarche-Gagnon G., Lay CY., Whyte L. (2013) Left Out in the Cold: Life in Cryoenvironments. In: Seckbach J., Oren A., Stan-Lotter H. (eds) Polyextremophiles. Cellular Origin, Life in Extreme Habitats and Astrobiology, vol 27. Springer, Dordrecht, p. 335–63.

[19] Bakermans C. Limits for microbial life at subzero temperatures. In: Margesin R, Schinner F, Marx JC, Gerday C, editors. Psychrophiles: from biodiversity to biotechnology. Berlin, Heidenberg: Springer; 2008. p. 17–28.

[20] Feller G, Gerday C. Psychrophilic enzymes: hot topics in cold adaptation. Nat Rev Microbiol. 2003;1(3):200–8.

[21] Rothschild LJ, Mancinelli RL. Life in extreme environments. Nature. 2001;409(6823):1092–101.

[22] Clarke A, Morris GJ, Fonseca F, Murray BJ, Acton E, Price HC. A low temperature limit for life on Earth. PLoS One 2013;8:e66207.

[23] McKay CP, Mykytczuk NC, Whyte LG. Life in ice on other worlds. In: Robert V. Miller, Lyle G. Whyte, editors. Polar microbiology: life in a deep freeze. American Society of Microbiology; ASM Press, Washington, DC. 2012. p. 312.

[24] Giudice AL, Fani R. Antimicrobial potential of cold-adapted bacteria and fungi from polar regions. In: Rampelotto P. editor. Biotechnology of extremophiles. Springer, Cham 2016. p. 83–115.

[25] Sánchez LA, Gómez FF, Delgado OD. Cold-adapted microorganisms as a source of new antimicrobials. Extremophiles. 2009;13(1):111–20.

[26] Chin JP, Megaw J, Magill CL, et al. Solutes determine the temperature windows for microbial survival and growth. Proceedings of the National Academy of Sciences 2010;107:7835–7840.

[27] Mykytczuk NC, Wilhelm RC, Whyte LG. *Planococcus halocryophilus* sp. nov., an extreme sub zero species from high Arctic permafrost. Int J Syst Evol Microbiol. 2012;62(8):1937–44.

[28] Mykytczuk NC, Foote SJ, Omelon CR, Southam G, Greer CW, Whyte LG. Bacterial growth at– 15 C; molecular insights from the permafrost bacterium Planococcus halocryophilus Or1. The ISME journal 2013;7:1211–26.

[29] Bakermans C, Skidmore ML. Microbial metabolism in ice and brine at –5 C. Environ Microbiol. 2011;13(8):2269–78.

[30] Fairén AG, Davila AF, Lim D, et al. Astrobiology through the ages of Mars: the study of terrestrial analogues to understand the habitability of Mars. Astrobiology 2010;10:821–43.

[31] McKay CP, Porco CC, Altheide T, Davis WL, Kral TA. The possible origin and persistence of life on Enceladus and detection of biomarkers in the plume. Astrobiology 2008;8:909–19.

[32] Marion GM, Fritsen CH, Eicken H, Payne MC. The search for life on Europa: limiting environmental factors, potential habitats, and Earth analogues. Astrobiology 2003;3:785–811.

[33] Tamppari L, Anderson R, Archer P, et al. Effects of extreme cold and aridity on soils and habitability: McMurdo Dry Valleys as an analogue for the Mars Phoenix landing site. Antarctic Science 2012;24:211–28.

[34] Pollard W, Haltigin T, Whyte L, et al. Overview of analogue science activities at the McGill Arctic Research Station, Axel Heiberg Island, Canadian High Arctic. Planetary and Space Science 2009;57:646–59.

[35] Pollard W, Omelon C, Andersen D, McKay C. Perennial spring occurrence in the Expedition Fiord area of western Axel Heiberg Island, Canadian high Arctic. Canadian Journal of Earth Sciences 1999;36:105–20.

[36] Cavicchioli R, Siddiqui KS, Andrews D, Sowers KR. Low-temperature extremophiles and their applications. Current Opinion in Biotechnology 2002;13:253–61.
[37] Liu J-T, Lu X-L, Liu X-Y, et al. Bioactive natural products from the antarctic and arctic organisms. Mini reviews in medicinal chemistry 2013;13:617–26.
[38] Adrio JL, Demain AL. Microbial enzymes: tools for biotechnological processes. Biomolecules. 2014;4(1):117–39.
[39] Bratchkova A, Ivanova V. Bioactive metabolites produced by microorganisms collected in Antarctica and the Arctic. Biotechnol Biotechnol Equip. 2011;25(suppl 1):1–7.
[40] Bruntner C, Binder T, Pathom-aree W, et al. Frigocyclinone, a novel angucyclinone antibiotic produced by a Streptomyces griseus strain from Antarctica. The Journal of antibiotics 2005;58:346–9.
[41] Mojib N, Philpott R, Huang JP, Niederweis M, Bej AK. Antimycobacterial activity in vitro of pigments isolated from Antarctic bacteria. Antonie van Leeuwenhoek 2010;98:531–40.
[42] Vester JK, Glaring MA, Stougaard P. Improved cultivation and metagenomics as new tools for bioprospecting in cold environments. Extremophiles. 2015;19(1):17–29.
[43] de Pascale D, De Santi C, Fu J, Landfald B. The microbial diversity of Polar environments is a fertile ground for bioprospecting. Marine genomics 2012;8:15–22.
[44] Kaeberlein T, Lewis K, Epstein SS. Isolating "uncultivable" microorganisms in pure culture in a simulated natural environment. Science. 2002;296(5570):1127–9.
[45] Bollmann A, Lewis K, Epstein SS. Incubation of environmental samples in a diffusion chamber increases the diversity of recovered isolates. Appl Environ Microbiol. 2007;73(20):6386–90.
[46] Bollmann A, Palumbo AV, Lewis K, Epstein SS. Isolation and physiology of bacteria from contaminated subsurface sediments. Applied and environmental microbiology 2010;76:7413–9.
[47] Pollock TJ. Gellan-related polysaccharides and the genus *Sphingomonas*. Microbiology. 1993;139(8):1939–45.
[48] Tamaki H, Hanada S, Sekiguchi Y, Tanaka Y, Kamagata Y. Effect of gelling agent on colony formation in solid cultivation of microbial community in lake sediment. Environmental microbiology 2009;11:1827–34.
[49] Mac Rygaard A, Thøgersen MS, Nielsen KF, Gram L, Bentzon-Tilia M. Effects of gelling agent and extracellular signaling molecules on the culturability of marine bacteria. Applied and Environmental Microbiology 2017;83:e00243–17.
[50] Gavrish E, Bollmann A, Epstein S, Lewis K. A trap for in situ cultivation of filamentous actinobacteria. Journal of microbiological methods 2008;72:257–62.
[51] Nichols D, Cahoon N, Trakhtenberg E, et al. Use of ichip for high-throughput in situ cultivation of "uncultivable" microbial species. Applied and environmental microbiology 2010;76:2445–50.
[52] Berdy B, Spoering AL, Ling LL, Epstein SS. In situ cultivation of previously uncultivable microorganisms using the ichip. Nature Protocols 2017;12:2232–42.
[53] Sizova M, Hohmann T, Hazen A, et al. New approaches for isolation of previously uncultivated oral bacteria. Applied and environmental microbiology 2012;78:194–203.
[54] Ling LL, Schneider T, Peoples AJ, et al. A new antibiotic kills pathogens without detectable resistance. Nature 2015;517:455–9.
[55] Goordial J, Altshuler I, Hindson K, Chan-Yam K, Marcolefas E, Whyte L. In situ field sequencing and life detection in remote (79° 26′ N) Canadian High Arctic permafrost ice wedge microbial communities. Frontiers in Microbiology 2017;8:2594.
[56] Aoi Y, Epstein S. New devices for cultivation, in manual of environmental microbiology. 4th ed. American Society of Microbiology; Washington DC, United States. 2016. p. 2.1. 3-1-2.1. 3–8.

[57] Ingham CJ, Sprenkels A, Bomer J, et al. The micro-Petri dish, a million-well growth chip for the culture and high-throughput screening of microorganisms. Proceedings of the National Academy of Sciences 2007;104:18217–22.
[58] Aoi Y, Kinoshita T, Hata T, Ohta H, Obokata H, Tsuneda S. Hollow-fiber membrane chamber as a device for in situ environmental cultivation. Applied and environmental microbiology 2009;75:3826–33.
[59] Ben-Dov E, Kramarsky-Winter E, Kushmaro A. An *in situ* method for cultivating microorganisms using a double encapsulation technique. FEMS Microbiol Ecol. 2009;68(3):363–71.
[60] Zengler K, Toledo G, Rappé M, et al. Cultivating the uncultured. Proceedings of the National Academy of Sciences 2002;99:15681–6.
[61] Dawoon Jung, Eun-Young Seo, Slava S. Epstein, Yochan Joung, Jaemin Han, Valentina V. Parfenova, Olga I. Belykh, Anna S. Gladkikh, Tae Seok Ahn, Application of a new cultivation technology, I-tip, for studying microbial diversity in freshwater sponges of Lake Baikal, Russia, *FEMS Microbiology Ecology*, Volume 90, Issue 2, November 2014, Pages 417–423, https://doi.org/10.1111/1574-6941.12399.
[62] Svenning MM, Wartiainen I, Hestnes AG, Binnerup SJ. Isolation of methane oxidising bacteria from soil by use of a soil substrate membrane system. FEMS microbiology ecology 2003;44:347–54.
[63] Ferrari BC, Winsley T, Gillings M, Binnerup S. Cultivating previously uncultured soil bacteria using a soil substrate membrane system. Nature protocols 2008;3:1261.
[64] Ferrari BC, Binnerup SJ, Gillings M. Microcolony cultivation on a soil substrate membrane system selects for previously uncultured soil bacteria. Appl Environ Microbiol. 2005;71(12):8714–20.
[65] Giavasis I, Harvey LM, McNeil B. Gellan gum. Crit Rev Biotechnol. 2000;20(3):177–211.
[66] Wüstenberg T. Cellulose and cellulose derivatives in the food industry: fundamentals and applications. John Wiley & Sons; Weinheim, Germany, 2014.
[67] Nobile M, Amendola E, Nicolais L, Acierno D, Carfagna C. Physical properties of blends of polycarbonate and a liquid crystalline copolyester. Polymer Engineering & Science 1989;29:244–57.
[68] Van Krevelen DW, Te Nijenhuis K. Properties of polymers: their correlation with chemical structure; their numerical estimation and prediction from additive group contributions. Elsevier; Amsterdam, The Netherlands, 2009.
[69] Kim KH, Gohtani S, Yamano Y. Effects of oil droplets on physical and sensory properties of o/w emulsion agar gel. J Texture Stud. 1996;27(6):655–70.
[70] Schmidt M, Priemé A, Stougaard P. Bacterial diversity in permanently cold and alkaline ikaite columns from Greenland. Extremophiles. 2006;10(6):551–62.
[71] Martinsen A, Skjåk-Bræk G, Smidsrød O. Alginate as immobilization material: I. Correlation between chemical and physical properties of alginate gel beads. Biotechnol Bioeng. 1989;33(1):79–89.

Yong-Hoe Choe and Yoo Kyung Lee

2 Analysis of lithic microbial communities

2.1 Introduction

2.1.1 Lithic habitats and their geology

Rock can be a sheltered habitat for microbes in the harsh environment [1], and the microbes living in this habitat are called as "lithic microbes". This lithic habitat affords a refuge from environmental stresses such as temperature, desiccation, and ultraviolet radiation, which make rocks as possible locations to find life in extreme environments or extraterrestrial habitats [2–7].

Lithic habitats are classified into six types depending on the place where organisms live on (epilithic), in (endolithic, cryptoendolithic, chasmoendolithic, euendolithic), or under the rock (hypolithic). The surface of the rocks offers an epilithic habitat, which serves as a stable environment for biofilm development. Compared to other lithic habitats, it is obvious that epilithic habitat is more sensitive to environmental changes [8–10]. The hypolithic habitat refers to the rock-soil interface at the lower surface of translucent rocks or small stones half-buried on the soil surface [11]. Most observations on hypolith are reported under translucent rocks such as quartz which allow transmission of light to support photosynthetic organisms [12, 13]. Endolithic communities inhabit the interstitial spaces of porous rocks [11]. Endolithic habitats offer various advantages with regard to environmental conditions. For example, the interior of the rock provides not only protection against strong ultraviolet radiation, but also the retained water in the pore of the rock can be a crucial source of water for endolithic microorganisms. In addition, the temperature fluctuations of endolithic habitat relatively less than the outside environment [14, 15].

According to previous studies, the endolithic habitat can be subdivided into cryptoendolithic, chasmoendolithic, and euendolithic habitat. The cryptoendolithic habitat is indirectly related to the rock surface and consists of pores of various sizes [16]. These habitats are mainly found in porous rocks such as sandstone. The chasmoendolithic communities colonize existing cracks or fissures of rocks. These habitats are mostly observed in weathered rocks. Euendolithic microorganisms, which bore into the rock by dissolving minerals, live deep in the rock and are commonly found in carbonate rocks [17].

Lithic microbial communities colonize a variety of rock substrates. However, the microbial community composition depends on the physical and chemical properties, such as porosity, water retention capacity, water permeability, and mineralogy of the rock [18–22]. The diverse lithic microorganisms are found in a variety of sedimentary rocks, such as sandstone [3, 4, 23], limestone [24, 25], gypsum [26–28], dolomite [29], carbonate [30], and halite [6]. In addition, the microbial colonization in rock

https://doi.org/10.1515/9783110525786-002

is also observed in igneous rocks, such as granite [2, 31], ignimbrite [7], and basalt [32]. Furthermore, several studies have conducted lithic communities associated with volcanic rocks [33–37]. These various rock substrates not only afford as reservoirs of diverse microorganisms but also serve as a refuge in extreme environments.

2.1.2 Distribution and diversity of lithic microorganisms

Several studies have indicated that the lithic microbial communities are relatively simple communities [38], but are observed in a wide range of environments. In the deep subsurface, the microbial activity is mainly driven by chemolithotrophs [39]. In contrast, lithic microbial communities at the surface of the Earth are dominated by either phototrophs and various bacterial heterotrophs, or by lichenizing fungi and their symbionts [38, 40, 41]. A small population of archaea is also observed in the lithic habitats.

Geographically, lithic microorganisms from a variety of extreme environments have been reported, including high-altitude in mountainous regions [24, 25], semi-arid areas [29], hyper-arid deserts [6, 7, 27], and cold deserts in the Arctic and Antarctic [2, 28]. Especially, a number of studies have been conducted in Antarctica Dry Valleys, where lithic habitat was considered to be a refuge for microorganisms. In the region, Cyanobacteria and lichen have been identified as dominant species [11, 42]. Cyanobacteria-dominated communities are composed of various microbial species, such as *Anabaena* sp. and *Chroococcidiopsis* sp. Among these microbial taxa, the cyanobacterium *Chroococcidiopsis* has widely distributed in lithic habitats of arid environments [13, 43, 44]. The possible reason for its wide distribution in such environments is due to their ability to resist strong UV and desiccation, as well as ability to do nitrogen fixation [45–47]. Lichen-dominated communities are composed of black or meristematic fungi associated with the algae [11, 48–50]. Friedmann [41] suggested that lithic lichens are regularly accompanied by bacteria, which presumably function as decomposers. These heterotrophic bacteria are actually found in association with lichen or cyanobacterial communities in almost all other lithic habitats [29, 51]. These heterotrophic bacteria belong to diverse bacterial groups such as *Actinobacteria*, *Acidobacteria*, and *Proteobacteria*.

2.1.3 Traditional approaches to assess lithic microbial community

To characterize lithic communities, most analyses have used traditional methods, including light microscopy and cultures. However, the limitations of such traditional methods have raised the need for new analytical techniques [52].

Culture-independent methods using metagenomic DNA-based data are now commonly used for microbial diversity assessments. Metagenomic analysis of lithic microbial communities can be performed in a variety of techniques, such as microbial fingerprinting, cloning for phylogenetically informative genes, and next-generation sequencing technologies. These molecular techniques are applied to identify lithic communities in several studies, such as the endolithic community in the Dry Valleys in Antarctica [53], the cyanobacterial community on the rocks in Switzerland [29], or the lithic microbial communities in the Rocky Mountains [1, 54, 55].

The development of DNA-sequencing technologies is expanding understanding of microbial genomes. Thus, knowledge of these techniques will become more important to understand microbial ecosystems. Moreover, lithic ecosystems are relatively simple compared to other microbial ecosystems, making them ideal for applying these technologies.

In this chapter, we introduce the basic steps of the research process, such as molecular tools and physiochemical analysis of rock, for analysis of lithic microbial communities.

2.2 Analysis methods for lithic microbial community

2.2.1 Rock sampling

Rock samples were collected using hardened steel chisels and geological hammer flame-sterilized in the field with 100% ethanol. Rock samples were placed in sterile plastic bags and transported to the laboratory in an icebox at ≤4°C. The rock samples were crushed using Mixer Mill, and then powdered rock samples were used to extract genomic DNA and to analyze the elemental composition of the rock.

2.2.2 X-ray fluorescence spectrometry (XRF) analysis

Major element composition of the rock samples was obtained using a wavelength-dispersive X-ray fluorescence spectrometry spectrometer (Panalytical, the Netherlands). The alkali flux used in this experiment was a lithium tetraborate ($Li_2B_4O_7$). The sample was ignited at 110°C for 8 h prior to weighing and cooled in a desiccator. Samples were then weighed in a ceramic crucible and ignited in a muffle furnace for 20 min/h at 1,000°C. Glass beads were prepared by mixing 0.6 g of powdered rock sample, or a mixture of rock sample and chemical reagents, and 6.0 g of the flux (10 times the amount of the standard powder sample). The mixture was put into a platinum-5% gold crucible, and 0.05 g of lithium bromide (LiBr) was added as an exfoliation agent. Fusing and agitation were carried out with an automated high-frequency bead sampler (SPEX, K-2 PRIME), at 300 s prefusion (~1,050°C), 60 s fusion (~1,080°C), and during agitation [56, 57].

2.2.3 Inductively coupled plasma-mass-spectrometer

Powdered samples were digested using HF and $HClO_4$ ($HF:HClO_4$=2:1). After washing with ultrasonic cleaning machine for 10 minutes, the sample solution was dried on a hot plate (120~180°C) and then dissolved with 7N HNO_3, and the supernatant was collected. This procedure from digestion to the supernatant collection was repeated at least three times. The final supernatant was dried on a hot plate (180°C) and then dissolved in 2% HNO_3. The concentration of trace elements was measured using inductively coupled plasma mass spectrometry (ICP-MS).

2.2.4 Mercury intrusion porosimetry (MIP)

To investigate porosity and pore size distribution, MIP test was performed using AutoPore IV (Micromeritics Instrument Corporation). In contrast to X-ray CT analysis, MIP method can analyze capillary pores (5 nm–10 μm). Moreover, because of its simplicity and ease, the MIP has been used to analyze the porosity and pore size of rock samples. However, this method is not suitable for studying the pore shape and structure of rocks.

2.2.5 Pore structure analysis by X-ray Microfocused Computed Tomography (X-ray μ-CT)

As a specimen for the X-ray CT analysis, the cylinders (5 mm diameter and 20 mm height) were obtained from rock samples to visualize internal pore distribution and network. To capture three-dimensional images of the specimens, X-ray CT was utilized using X-EYE system (Korea Institute of Civil Engineering and Building Technology): the system provides high-resolution images. Because pore distribution and structure analysis using X-ray CT is an image-based technique, the detectable minimum size of pore and the precision of the method depend on the resolution of the images. In contrast to the MIP method, X-ray CT analysis is suitable for detecting relatively large-sized pores. In addition, the advantage of the X-ray CT method is that it can be analyzed without destroying the sample, and the 3D image helps to intuitively understand the pore structure.

2.2.6 PCR amplification and sequencing

We extracted total genomic DNA extraction from rock samples according to the previous report [58]. PCR amplifications were carried out with primers pairs targeting the V3–V4 region of the bacterial *16S rRNA* gene and the D1–D2 region of the fungal

28S rRNA gene. The sequencing of *16S rRNA* gene amplicon was performed by ChunLab Inc. (Seoul, Korea) using 2 × 300 bp Miseq (Illumina) runs. Raw reads were assembled using PANDAseq [59], and were further processed following the MiSeq SOP using mothur [60]. The sequencing of *28S rRNA* gene amplicon was carried out by Macrogen Inc. (Seoul, Korea) using the 454 GS FLX Titanium Sequencing system (Roche 45 Life Science, USA). Raw sequences were initially filtered out low-quality sequences using PyroTrimmer [61]. The filtered sequences were further processed following the 454 SOP using mothur [62].

The taxonomic analysis of the resultant sequences was performed using the EzTaxon-e database for bacteria [63] and the fungal large subunit reference data at ribosomal database project (RDP) for fungi. We defined operational taxonomic units (OTUs) at 97% sequence similarity level.

2.2.7 Statistical analysis

Statistical analysis of sequence data is important to understand the microbiome. Diversity indices such as Simpson, Shannon, Ace, and Chao1 generated based on OTUs allow understanding the evenness and richness of the microbial communities [64–65]. Several ordination methods such as PCoA and nMDS are able to view the relationship between microbial communities.

Principal Coordinate Analysis (PCoA) was useful for microbial community analysis because we could use phylogenetic distances and community composition measures to calculate similarities between microbial populations. However, PCoA cannot directly associate the measured variables with individual principal coordinate axes because it used a distance matrix as its input [66]. Instead, PCoA can be used to evaluate the contribution of the variable to object dispersion along a specific Principal Coordinate axis [67]. NMDS (nonmetric multidimensional scaling) was also useful for microbial community analysis. Similar to PCoA, NMDS first calculated the matrix of object dissimilarities using the selected distance metric, and then ranks of the distances among the objects were calculated. Then it represented a configuration of objects, which best matches differences in ranks, in the selected N-dimensional ordination space [68].

In the generated ordination, sample proximity represented relative similarity. Closer communities were more similar than communities that were located further apart in the ordination. In terms of the quality of the ordination, the stress value can be used to assess the accuracy of the ordination. Stress values below 0.15 were considered commonly acceptable [69]. In addition, multivariate statistical tests such as Mantel test, ANOSIM, and PERMANOVA are able to quantify differences among communities after assigning the communities to groups. Mantel test was used to determine the correlation between two distance matrices (e.g. OTUs dissimilarity matrix and environmental factor matrix) [70]. The Analysis of Similarities (ANOSIM) tests for a significant difference between two or more groups was also used. We compared the ranks of distances between groups with ranks of distance within groups [69]. PERMANOVA (Permutational Multivariate Analysis of Variance) was used to test for significant differences between two or more groups [71].

To determine that certain OTUs represent more in certain habitat (e.g. rock type), OTU abundance and taxonomy information were analyzed using DESeq2 or edgeR [64, 72].

By using the various statistical methods described above, the ecological characteristics of lithic microbial communities can be understood more clearly.

2.3 Concluding remarks

The abundance, phylogenetic diversity, and variability of microbial communities in lithic environments indicate that rock can be one of important habitat for life, particularly in extreme environments. In other words, a study on lithic microorganisms can provide insight into the concept that how organisms can survive in extreme environments. These insights will not only expand our knowledge of the microbial ecosystem on earth but also help to find the extraterrestrial life in the field of astrobiology.

Acknowledgments: This work was supported by a grant from Korea Polar Research Institute (PE16030 and PE17280).

References

[1] Walker JJ, Pace NR. Phylogenetic composition of Rocky Mountain endolithic microbial ecosystems. Applied and Environmental Microbiology 2007;73:3497–504.

[2] De los Rios A, Sancho LG, Grube M, Wierzchos J, Ascaso C. Endolithic growth of two Lecidea lichens in granite from continental Antarctica detected by molecular and microscopy techniques. New Phytologist 2005;165:181–9.

[3] Friedmann EI. Endolithic microbial life in hot and cold deserts. Orig Life 1980;10:223–35.

[4] Friedmann EI, Kappen L, Meyer MA, Nienow JA. Long-Term Productivity in the Cryptoendolithic Microbial Community of the Ross Desert, Antarctica. Microbial Ecology 1993;25:51–69.

[5] Parnell J, Lee P, Cockell CS, Osinski GR. Microbial colonization in impact-generated hydrothermal sulphate deposits, Haughton impact structure, and implications for sulphates on Mars. Int J Astrobiol 2004;3:247–56.

[6] Wierzchos J, Ascaso C, McKay CP. Endolithic cyanobacteria in halite rocks from the hyperarid core of the Atacama Desert. Astrobiology 2006;6:415–22.

[7] Wierzchos J, Davila AF, Artieda O, et al. Ignimbrite as a substrate for endolithic life in the hyper-arid Atacama Desert: Implications for the search for life on Mars. Icarus 2013;224:334–46.

[8] Cameron RE. Farthest south algae and associated bacteria. Phycologia 1972;11:133–9.

[9] Engelskjøn T. Botany of two Antarctic mountain ranges: Gjelsvikfjella and Muhlig-Hofmannfjella, Dronning Maud Land. Polar Research 1986;4:205–24.

[10] Parker B, Ford A, Allnutt T, Bishop B, Wendt S. Baseline microbiological data for soils of Dufek Massif. Antarctic Journal of the United States 1977;12:24–6.
[11] Nienow JA, Friedmann EI. Terrestrial lithophytic (rock) communities. Antarctic microbiology 1993:343–412.
[12] Broady PA. The ecology of sublithic terrestrial algae at the Vestfold Hills, Antarctica. British phycological journal 1981;16:231–40.
[13] Schlesinger WH, Pippen JS, Wallenstein MD, Hofmockel KS, Klepeis DM, Mahall BE. Community composition and photosynthesis by photoautotrophs under quartz pebbles, southern Mojave Desert. Ecology 2003;84:3222–31.
[14] Cockell CS, McKay CP, Omelon C. Polar endoliths–an anti-correlation of climatic extremes and microbial biodiversity. Int J Astrobiol 2002;1:305–10.
[15] Cockell CS, Lee P. The biology of impact craters–a review. Biological Reviews 2002;77:279–310.
[16] Saiz-Jimenez C, Garcia-Rowe J, Del Cura MG, Ortega-Calvo J, Roekens E, Van Grieken R. Endolithic cyanobacteria in Maastricht limestone. Science of the Total Environment 1990;94:209–20.
[17] Cockell CS, Lee P, Broady P, et al. Effects of asteroid and comet impacts on habitats for lithophytic organisms—A synthesis. Meteoritics & Planetary Science 2005;40:1901–14.
[18] Cockell CS, Olsson-Francis K, Herrera A, Meunier A. Alteration textures in terrestrial volcanic glass and the associated bacterial community. Geobiology 2009;7:50–65.
[19] Cockell CS, Olsson K, Knowles F, et al. Bacteria in weathered basaltic glass, Iceland. Geomicrobiology Journal 2009;26:491–507.
[20] Herrera A, Cockell CS, Self S, et al. A cryptoendolithic community in volcanic glass. Astrobiology 2009;9:369–81.
[21] Kelly LC, Cockell CS, Herrera-Belaroussi A, et al. Bacterial diversity of terrestrial crystalline volcanic rocks, Iceland. Microbial ecology 2011;62:69–79.
[22] Omelon CR, Pollard WH, Ferris FG. Inorganic species distribution and microbial diversity within high Arctic cryptoendolithic habitats. Microbial ecology 2007;54:740–52.
[23] Omelon CR, Pollard WH, Ferris FG. Environmental controls on microbial colonization of high Arctic cryptoendolithic habitats. Polar Biology 2006;30:19–29.
[24] Ferris FG, Lowson EA. Ultrastructure and geochemistry of endolithic microorganisms in limestone of the Niagara Escarpment. Canadian Journal of Microbiology 1997;43:211–9.
[25] Wong FKY, Lau MCY, Lacap DC, Aitchison JC, Cowan DA, Pointing SB. Endolithic Microbial Colonization of Limestone in a High-altitude Arid Environment. Microbial Ecology 2010;59:689–99.
[26] Boison G, Mergel A, Jolkver H, Bothe H. Bacterial life and dinitrogen fixation at a gypsum rock. Applied and environmental microbiology 2004;70:7070–7.
[27] Dong HL, Rech JA, Jiang HC, Sun H, Buck BJ. Endolithic cyanobacteria in soil gypsum: Occurrences in Atacama (Chile), Mojave (United States), and Al-Jafr Basin (Jordan) deserts. Journal of Geophysical Research-Biogeosciences 2007;112.
[28] Hughes KA, Lawley B. A novel Antarctic microbial endolithic community within gypsum crusts. Environmental Microbiology 2003;5:555–65.
[29] Sigler WV, Bachofen R, Zeyer J. Molecular characterization of endolithic cyanobacteria inhabiting exposed dolomite in central Switzerland. Environmental Microbiology 2003;5:618–27.
[30] Hoppert M, Flies C, Pohl W, Günzl B, Schneider J. Colonization strategies of lithobiontic microorganisms on carbonate rocks. Environmental Geology 2004;46:421–8.
[31] Gleeson DB, Clipson N, Melville K, Gadd GM, McDermott FP. Characterization of fungal community structure on a weathered pegmatitic granite. Microbial Ecology 2005;50:360.
[32] Izawa M, Banerjee N, Flemming R, Bridge N, Schultz C. Basaltic glass as a habitat for microbial life: implications for astrobiology and planetary exploration. Planetary and Space Science 2010;58:583–91.

[33] Costello EK, Halloy SRP, Reed SC, Sowell P, Schmidt SK. Fumarole-Supported Islands of Biodiversity within a Hyperarid, High-Elevation Landscape on Socompa Volcano, Puna de Atacama, Andes. Applied and Environmental Microbiology 2009;75:735–47.
[34] Dunfield KE, King GM. Molecular analysis of carbon monoxide-oxidizing bacteria associated with recent Hawaiian volcanic deposits. Applied and Environmental Microbiology 2004;70:4242–8.
[35] Gomez-Alvarez V, King GM, Nusslein K. Comparative bacterial diversity in recent Hawaiian volcanic deposits of different ages. Fems Microbiology Ecology 2007;60:60–73.
[36] Herrera A, Cockell CS, Self S, et al. Bacterial colonization and weathering of terrestrial obsidian in Iceland. Geomicrobiology Journal 2008;25:25–37.
[37] King GM. Contributions of atmospheric CO and hydrogen uptake to microbial dynamics on recent Hawaiian volcanic deposits. Applied and Environmental Microbiology 2003;69:4067–75.
[38] Walker JJ, Pace NR. Endolithic microbial ecosystems. Annu Rev Microbiol. 2007;61:331–47.
[39] Lovley DR, Chapelle FH. Deep subsurface microbial processes. Reviews of Geophysics 1995;33:365–81.
[40] Pointing SB, Chan Y, Lacap DC, Lau MC, Jurgens JA, Farrell RL. Highly specialized microbial diversity in hyper-arid polar desert. Proceedings of the National Academy of Sciences 2009;106:19964–9.
[41] Friedmann EI. Endolithic microorganisms in the Antarctic cold desert. Science 1982;215:1045–53.
[42] De Los Ríos A, Wierzchos J, Ascaso C. The lithic microbial ecosystems of Antarctica's McMurdo Dry Valleys. Antarctic Science 2014;26:459–77.
[43] Smith MC, Bowman JP, Scott FJ, Line MA. Sublithic bacteria associated with Antarctic quartz stones. Antarctic Science 2000;12:177–84.
[44] Warren-Rhodes KA, Rhodes KL, Pointing SB, et al. Hypolithic cyanobacteria, dry limit of photosynthesis, and microbial ecology in the hyperarid Atacama Desert. Microbial Ecology 2006;52:389–98.
[45] Billi D, Friedmann EI, Hofer KG, Caiola MG, Ocampo-Friedmann R. Ionizing-radiation resistance in the desiccation-tolerant cyanobacteriumChroococcidiopsis. Applied and Environmental Microbiology 2000;66:1489–92.
[46] Potts M. Mechanisms of desiccation tolerance in cyanobacteria. European Journal of Phycology 1999;34:319–28.
[47] Fewer D, Friedl T, Büdel B. Chroococcidiopsis and heterocyst-differentiating cyanobacteria are each other's closest living relatives. Molecular phylogenetics and evolution 2002;23:82–90.
[48] Friedmann EI, Hua M, Ocampo-Friedmann R. 3.6 Cryptoendolithic lichen and cyanobacterial communities of the Ross Desert, Antarctica. Polarforschung 1988;58:251–9.
[49] Selbmann L, De Hoog G, Mazzaglia A, Friedmann E, Onofri S. Fungi at the edge of life: cryptoendolithic black fungi from Antarctic desert. Stud Mycol 2005;51:1–32.
[50] Ruisi S, Barreca D, Selbmann L, Zucconi L, Onofri S. Fungi in Antarctica. Reviews in Environmental Science and Bio/Technology 2007;6:127–41.
[51] Pointing SB, Warren-Rhodes KA, Lacap DC, Rhodes KL, McKay CP. Hypolithic community shifts occur as a result of liquid water availability along environmental gradients in China's hot and cold hyperarid deserts. Environmental microbiology 2007;9:414–24.
[52] Pace NR. A molecular view of microbial diversity and the biosphere. Science 276: 734–740. 1997.
[53] José R, Goebel BM, Friedmann EI, Pace NR. Microbial diversity of cryptoendolithic communities from the McMurdo Dry Valleys, Antarctica. Applied and Environmental Microbiology 2003;69:3858–67.
[54] Garcia-Pichel F, López-Cortés A, Nübel U. Phylogenetic and morphological diversity of cyanobacteria in soil desert crusts from the Colorado Plateau. Applied and Environmental Microbiology 2001;67:1902–10.
[55] Norris TB, Castenholz RW. Endolithic photosynthetic communities within ancient and recent travertine deposits in Yellowstone National Park. FEMS microbiology ecology 2006;57:470–83.

[56] Fitton JG, Mahoney JJ, Wallace PJ, Saunders AD. Origin and Evolution of the Ontong Java Plateau. Geological Society, London, Special Publications. 2004;229:1–8.
[57] Mori Y, Mashima H. X-ray fluorescence analysis of major and trace elements in silicate rocks using 1: 5 dilution glass beads. Bull Kitakyushu Mus Nat Hist Hum History Series A. 2005;3:1–12.
[58] Choe Y-H, Kim M, Woo J, Lee MJ, Lee JI, Lee EJ, Lee YK. Comparing Rock-inhabiting Microbial Communities in Different Rock Types from a High Arctic Polar Desert. FEMS Microbiology Ecology 2018;94: fiy070.
[59] Masella AP, Bartram AK, Truszkowski JM, Brown DG, Neufeld JD. PANDAseq: paired-end assembler for illumina sequences. BMC bioinformatics 2012;13:31.
[60] Koch J, Günther D. Review of the state-of-the-art of laser ablation inductively coupled plasma mass spectrometry. Applied Spectroscopy 2011;65:155A-62A.
[61] Oh J, Kim BK, Cho W-S, Hong SG, Kim KM. PyroTrimmer: a software with GUI for pre-processing 454 amplicon sequences. J Microbiol. 2012;50:766–9.
[62] Schloss PD, Westcott SL, Ryabin T, et al. Introducing mothur: open-source, platform-independent, community-supported software for describing and comparing microbial communities. Applied and environmental microbiology 2009;75:7537–41.
[63] Kim O-S, Cho Y-J, Lee K, et al. Introducing EzTaxon-e: a prokaryotic 16S rRNA gene sequence database with phylotypes that represent uncultured species. Int J Syst Evol Microbiol. 2012;62:716–21.
[64] Anders S, Huber W. Differential expression analysis for sequence count data. Genome biology 2010;11:R106.
[65] McMurdie PJ, Holmes S. phyloseq: an R package for reproducible interactive analysis and graphics of microbiome census data. PloS one 2013;8:e61217.
[66] Ramette A (2007) Multivariate analyses in microbial ecology. FEMS Microbiology Ecology, 62, 142–160.
[67] Koenig JE, Spor A, Scalfone N et al. (2011) Succession of microbial consortia in the developing infant gut microbiome. Proceedings of the National Academy of Sciences, USA, 108(Suppl 1), 4578–4585.
[68] Kruskal JB (1964) Multidimensional scaling by optimizing goodness of fit to a nonmetric hypothesis. Psychometrika, 29, 1–27.
[69] Clarke KR (1993) Nonparametric multivariate analyses of changes in community structure. Australian Journal of Ecology, 18, 117–143.
[70] Mantel N (1967) The detection of disease clustering and a generalized regression approach. Cancer Research, 27, 209–220.
[71] Anderson MJ (2001) A new method for non-parametric multivariate analysis of variance. Austral Ecology, 26, 32–46.
[72] McCarthy DJ, Chen Y, Smyth GK (2012) Differential expression analysis of multifactor RNA-seq experiments with respect to biological variation. Nucleic Acids Res, 40:4288–4297.

Júnia Schultz and Alexandre Soares Rosado

3 Use of microbes from extreme environments for biotechnological applications

3.1 Introduction

Recent years have seen the discovery of many organisms that are capable of inhabiting extreme environments considered lethal to humans and other forms of life, increasingly expanding the known limits of living beings [1] (Fig. 3.1). These organisms are called extremophiles, and they not only tolerate environmental extremes but also require these conditions for optimal growth and development [2, 3]. Extremophiles are members of all three domains of life, but include mainly prokaryotic organisms, i.e. Bacteria and Archaea, and a few eukaryotes such as microalgae, fungi, and protozoa, which also possess adaptations to certain extreme environmental conditions [3, 4].

Extremophiles are classified according to the physical and chemical conditions in their habitats: (1) psychrophilic, thermophilic, and hyperthermophilic, which grow under conditions of very low, high, and very high temperatures, respectively

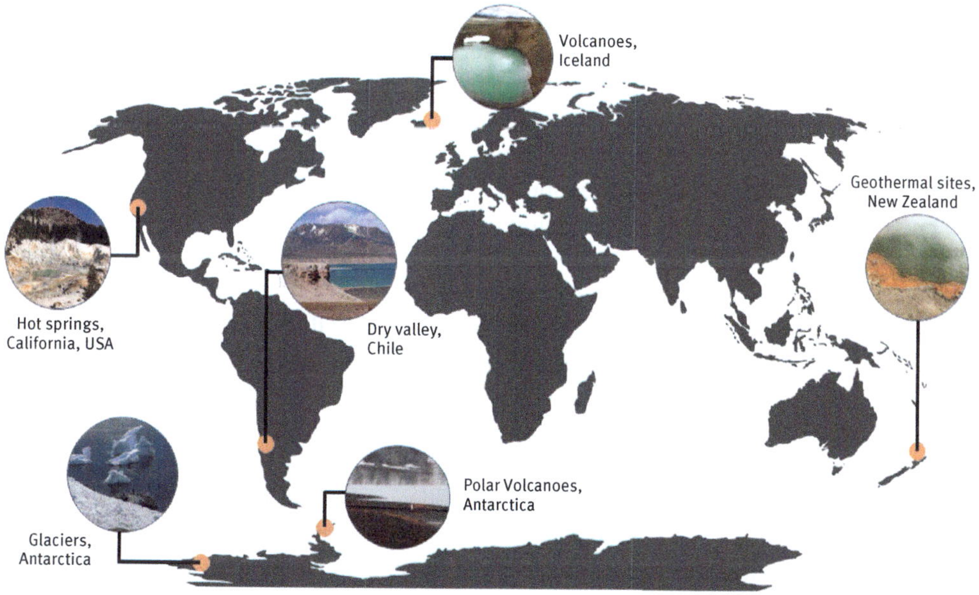

Fig. 3.1: Different extreme environments present around the Earth, such as cold environments in Antarctica and Arctic, hot springs in California, USA, and geothermal sites in New Zealand, where extremophiles can be found. Source: Authors.

https://doi.org/10.1515/9783110525786-003

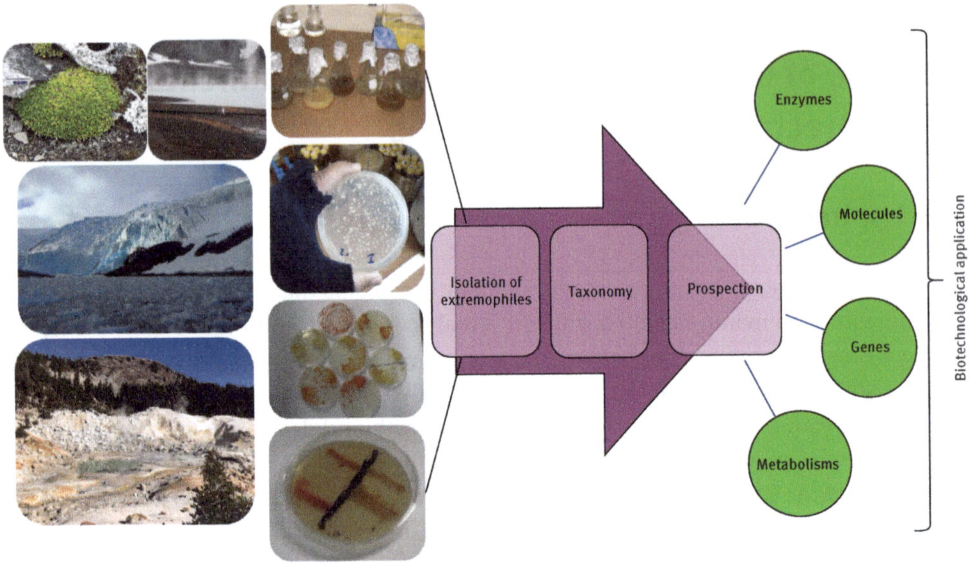

Fig. 3.2: Bioproducts, such as enzymes, molecules, genes, and metabolisms, offered by microorganisms from extreme environments of industrial interest, for example, application in pharmaceutical, medical and cosmetics industries, bioenergy and biofuel companies, food and feed industry, and bioremediation processes. Source: Authors.

[5]; (2) acidophilic and alkaliphilic, which grow at low and high pH levels, respectively [6, 7]; (3) piezophilic, which grow under high pressure [8]; (4) halophilic, which require high salt concentrations [9]; (5) osmophilic, which tolerate high osmotic pressure [10]; (6) xerophilic, which tolerate low water availability [11]; (7) oligotrophic, which require low concentrations of nutrients [2]; and (8) radioresistant, which tolerate a high incidence of radiation [12]. Organisms that develop in more than one of these extreme conditions, living under multiple forms of stress, are called polyextremophiles [13].

Being present in extreme environments, these organisms undergo selective pressure in order to meet the constant environmental challenges and develop adaptations for local survival, including unique and versatile metabolic capacities [14]. These characteristics make extremophiles good sources of new bioproducts for use in the most diverse areas of biotechnology (Fig. 3.2), including processes performed under harsh conditions, or to replace chemicals with sustainable processes [15, 16], including the petroleum, bioenergy, pharmaceutical, cosmetic, food, pulp and paper, molecular genetics, detergent, and leather tanning industries [14, 17–20].

This review describes some groups of the extremophilic microorganisms known from different extreme environments; discusses how these and/or their products can be applied in the different areas of biotechnology, based on recent literature; and discusses current advances and proposed directions for new studies on the use of microorganisms from extreme environments for industrial applications.

3.2 Extremophilic microbe potential: the search for bioproducts and their biotechnological applications

Since the earliest discoveries of microorganisms exploiting these different extreme environments, interest in the biotechnological potential of this material has steadily grown. Extremophiles have become the main targets in the search and discovery of new natural products, new biocatalysts, and other goods and services that they may provide. Applications of these extreme microorganisms have opened an exciting phase of biotechnology innovation and have been increasing with the advent of new biological technologies.

The following subtopics illustrate the biotechnological potential of some groups of extremophilic microorganisms.

3.2.1 Thermophiles and hyperthermophiles

Thermophilic and hyperthermophilic microorganisms are defined as organisms that not only survive but also thrive at high temperatures with thermophiles requiring temperatures over 45°C and hyperthermophiles reaching their optimum development at temperatures over 85°C [21]. Until recently, microbial life was known only in temperatures up to 113°C, such as the chemolithotrophic archaeon *Pyrolobus fumari*. However, studies with microorganisms surviving at even higher temperatures are continually being reported; Takai et al. [22] described the methanogenic hyperthermophilic archaeon *Methanopyrus kandleri*, which grows at 122°C and is the record holder for thermophilic microorganisms.

The vast majority of hyperthermophilic species belong to the Domain Archaea, but thermophiles are distributed throughout the three domains of life, especially in the Bacteria and Archaea domains [2]. Within the Bacteria, *Aquifex pyrophilus* and *Thermotoga maritima* can upkeep growth at the highest temperatures, 95 and 90°C, respectively [23]. These microorganisms have been isolated from the most varied natural thermophilic environments, including volcanoes, terrestrial hot springs, deep-sea hydrothermal vents, deserts, coal piles, compost heaps, and oil reservoirs [24–26], as well as in artificial environments, for instance domestic and commercial wastewater treatment systems, compost, and biological reactors [27–29].

To survive in these environments without allowing high temperatures to affect their development, thermophiles and hyperthermophiles have developed a number of cellular- and molecular-level adaptations [30]. These include changes in the cell membrane, where the Archaea possess modified lipids, such as a lipid monolayer, and thermophiles and hyperthermophiles have higher levels of saturated fatty acids, all of which increase the rigidity and stability of the plasma membrane to protect against high temperatures [31, 32]. Mandelli et al. [33, 34] observed that carotenoids play an important role in the thermal stability of microbial structures and in the thermophilic bacterium *Thermus filiformis* (ATCC 43280) have antioxidant properties.

According to Hara et al. [35], carotenoids can be inserted into lipid bilayers in order to strengthen the membrane for biotechnological applications at high temperatures.

Biological adaptations are present at the molecular level as well, as these extreme conditions may rupture the hydrogen bonds of the DNA molecule, denaturing it. Some hyperthermophilic Archaea have developed thermal resistance in the double strand of DNA, which is improved by the enzyme reverse gyrase that causes overlapping folding of DNA, considerably increasing its thermal stability [36].

Another mechanism for microbial adaptation at high temperatures, according to van Wolferen et al. [37], is horizontal gene transfer, which functions to repair DNA damage by obtaining new genes that can aid microbes to survive in extreme conditions. RNA molecules are also affected by high temperatures, and to prevent denaturation, RNA secondary structures are stabilized by increasing the guanine/cytosine content and adding post-transcriptional modifications [38].

High temperatures also destabilize the protein structures of the cells. One strategy to increase stability is the presence of hydrophobic residues within their structures, which interact with the protein core [39]. Another strategy is the presence of heat-shock proteins, which are expressed by all cells in response to environmental stress conditions, where they perform chaperone functions, stabilizing new proteins and ensuring that they fold correctly [40]. Large et al. [41] found that at 108°C, the hyperthermophilic *Pyrodictium occultum* contained approximately 80% soluble proteins called thermosomes, which are heat-induced chaperone proteins.

Because of these structural and metabolic adaptations, to remain active and stable at high temperatures, thermophiles and hyperthermophiles have been intensely studied in the search for new genes, metabolites, and enzymes for different industrial uses. These thermophilic bioproducts are highly appropriate for use in the emerging industrial biotechnology [14].

Thermophilic and hyperthermophilic proteins are important, since thermostatic enzymes, also called thermoenzymes, have gained in importance in biotechnological processes in recent years. Enzymes are essential targets for thermal adaptation, as they catalyze all chemical reactions that keep organisms alive. Little is known about the enzymatic adaptations of thermophiles, except that they show a relationship between stiffness and flexibility, making it difficult to interact with the substrate but increasing enzyme specificity [42].

From the industrial point of view, hyperthermophilic enzymes have advantages over mesophilic enzymes because their capacity to remain active and efficient at high temperatures enables them to remain stable; execute faster reactions; increase the yield of a process by increasing the solubility of the substrate and reducing the feed time; reduce the risk of contamination by other microorganisms and viruses, which could inactivate the enzyme; and be separated more easily in the purification process [19, 43]. They also exhibit resistance to denaturing agents, such as detergents and organic solvents, conditions in which their mesophilic counterparts are inactive, and are commonly used to solubilize substrates of interest. These properties have attracted the attention of the biotechnology industry [44–46].

For these reasons, the scientific community has been seeking alternatives to traditional methods, using enzymes from extreme thermophiles in order to optimize industrial production and reduce process costs, as well as to make use of sustainable and non-chemical products [47].

The following studies described the use of the enzymes lipase [48], laccase [49], superoxide dismutase [50], amylase [51], esterase [52], protease [53], cellulase [54], and xylanase [55]. Polymerase and thermostable oxidases are still being studied and have numerous applications in different industries, including food processing, leather tanning, detergents, paper and pulp, textiles, biofuels, pharmaceuticals, chemicals, and bioremediation [19, 56–57]. According to Egorova and Antranikian [58], the thermophilic genera that are most often studied for enzyme production are *Bacillus*, *Thermus*, *Thermotoga*, *Thermococcus*, *Pyrococcus*, and *Sulfolobus*, obtained from different environments such as volcanoes, geothermal environments, and deep-water hydrothermal vents. Examples include the esterase produced by *Pyrococcus abyssi*, which maintains optimal activity at 90°C and a half-life of thermal inactivation at 99°C after 22 h of incubation [59]; *Rhodothermus marinus*, which produces thermostable amylase with a maximum activity at 80°C [60]; and *Thermus aquaticus* and *Pyrococcus furiosus*, which produce Taq polymerase arrays and Pfu polymerase, respectively, which are widely used in molecular biology research [61]. Thermophilic cellulases are widely studied, especially for the production of second-generation ethanol. In the hot springs of Yellowstone National Park alone, more than 40 thermoenzymes have been found and have already been implemented in food processing and bioethanol production [20].

3.2.2 Psychrophiles

The coldest habitats include glaciers, alpine regions, snow, profundal ocean environments, the polar oceans, ice caves, and permafrost [62] and represent more than 90% of the total biosphere [63]. These apparently inhospitable environments harbor a wide diversity of microorganisms, including bacteria, archaea, photoautotrophic cyanobacteria and green algae, fungi, and protozoa, which successfully colonized them by means of adaptations to the cold [64].

These organisms, belonging to all three domains of life, are termed psychrophiles, defined as having optimum growth temperatures below 15°C and failure to grow at temperatures above 20°C [65]. Psychrophiles are the most abundant extremophiles on Earth in terms of biomass, diversity, and distribution [16, 66]. Bacteria are the most common members of this group [62].

Active psychrophilic microorganisms have been recorded at temperatures as low as −20°C [5]. In order to overcome the various challenges imposed by physical stress from cold, such as the structural destruction of cells by ice crystals, interactions of modified macromolecules, and denaturation of biomolecules, psychrophiles have developed a series of strategies [2, 5].

One of the negative impacts of low temperatures is the reduction of membrane fluidity. Some adaptations are the increase in unsaturated fatty acids in phospholipids, change in lipid composition, reduction in lipid size and load, and conversion of fatty acids from -trans to -cis. Additionally, the membrane may produce polar and nonpolar carotenoids, as observed in several Antarctic bacteria, and these pigments help to modulate membrane fluidity and maintain viscosity [5].

Another impact on microbial cells is the formation of ice crystals in the cytoplasm, which results in cell damage and osmotic imbalance. Some cryoprotection strategies include the accumulation of compatible solutes such as glycine, betaine, sucrose, and mannitol, which lower the cytoplasmic freezing point, providing protection from freezing as well as from desiccation and hyperosmolarity [67]. Some psychrophilic species produce antifreeze proteins, which bind and control the growth of ice crystals and recrystallization, again lowering the freezing point. Others produce ice-nucleating proteins that control and stabilize the formation of ice crystals, helping to prevent membrane rupture in low temperatures [68].

The protein structures of psychrophiles contain less alanine, arginine, and proline, amino acids that reduce conformational flexibility. In contrast, the proteins contain more of the amino acids asparagine, methionine, and glycine, compounds that increase the local mobility of protein, reduce the number of hydrogen bonds and salt bridges, and consequently increase protein flexibility [69, 70]. Additionally, an increase in hydrophobic amino acids was observed on the exposed side of the proteins, and fewer hydrophobic residues were found in their centers [5].

In general, psychrophilic genomes show a high level of redundancy; they contain multiple copies of transporter RNA, which is related to amino acid synthesis, as well as an increased variety and number of chaperones [71]. The genome of these organisms possesses a large number of features involved with genomic plasticity, such as plasmids, transposons, and other mobile genetic elements, where the genes that they carry are directly involved in adaptations to cold, such as acid biosynthesis [72].

Among the various effects of low temperatures is the slowing of chemical reactions, and because of this, the enzymes of psychrophiles must also be adapted to maintain suitable catalytic rates for cellular functions. According to Cavicchioli et al. [73], psychrophilic enzymes are generally characterized by highly flexible structures, low thermostability, and high specificity at low temperatures, which reduce their activation energy and increase substrate turnover rates.

Psychrophilic enzymes are sought after and have many areas of application due to industrial factors, such as efficient production at low cost, saving of energy, and importance for thermal protection, which increases the quality of different products. Because of these characteristics, the psychrophiles have also attracted the attention of researchers and industries due to their biotechnological potential [16]. Cold-adapted enzymes such as cellulases and psychrophilic chitinases are used in energy production and biofuel processing; and peroxidases, lipases, and hydrocarbon-degrading enzymes capable of activity at low temperatures are used in bioremediation processes

[74, 75]. Another area that requires the use of cold-adapted enzymes is pharmaceutics and medicine, especially chitinase, lipase, laccase, lactamase, peroxidase, and esterase [76]. Lipases and esterases are widely used in the detergent and cleaning-products industries; and lipase, protease, glucanase and xylanase find uses in the food and feed industries [77–79]. An example of the benefits of cold-adapted microorganisms is their use in lactose hydrolysis at low temperatures in the milk-storage process. This bioprocess was patented by Nutrilab NV in Belgium and is used to produce milk for patients with lactose intolerance [62].

3.2.3 Acidophiles

Microorganisms that grow in highly acidic environments, with pH levels below 3, are termed acidophiles and can be found in all three domains of life, Archaea, Bacteria, and Eukarya [7, 80], but occur most widely in the Bacteria and Archaea [81]. They have an optimal pH between 2 and 4; microorganisms capable of growing at pH levels close to 0 have been described, such as the archaeons *Thermoplasma acidophilum, Thermoplasma volcanium*, and *Picrophilus oshimae* [82].

Acidic habitats are abundant on Earth. An array of natural processes cause acidity in the environment, including volcanic soils, fumaroles, sulfur fields, sulfur pools, geysers, mineral-rich environments, and hot springs. Most low-pH environments are at least partially anthropogenic, such as mining activities, industrial leaching, and coal and waste heaps [7, 83, 84].

In order to thrive in these environments, acidophiles have developed cellular adaptations to overcome the ecophysiological challenges posed by low pH, but they especially needed efficient mechanisms to maintain the pH of the cytoplasm at or close to neutral, as acidophiles can survive in extremely high-acid environments but do not tolerate such conditions within the cell, due to macromolecular instability [85].

The first barrier, the cell membrane, differs in fatty acid and lipid composition and contains lipid tetraethers that are less susceptible to acid hydrolysis [86, 87]. Other processes involved in maintaining internal homeostasis of acidophilic microorganisms have been reported by Baker-Austin and Dopson [85], including reversal of the membrane potential to partially divert the inner flow of protons, cell membranes that are highly impermeable in order to retard the influx of protons into the cell, and a cytoplasmic buffer.

Genomic analyses have found a higher proportion of secondary carriers in the genomes of acidophiles than neutrophiles, which reduces the energy demand for pumping solutes and nutrients to the cell, as well as a higher proportion of DNA and protein-repair systems in acidophiles. Other mechanisms include minimizing solvents and binding to metal cofactors [85, 88].

The acidophilic proteins are also adapted to remain stable and active in the acidic environment [89], for example, pepsin in *Sulfolobus acidocaldarius*. Acidophilic enzymes are rich in acid residues on their surface, such as endo-β-glucanase in the archaeon *Sulfolobus solfataricus*, which is active at pH 1.8 and tolerates a high concentration of hydrogen ions [90]. Another example, described by Michaux et al. [91], is endo-1,4-β-xylanase in the fungus *Scytalidium acidophilum*, which shows structural modifications compared to its homologue, including alteration in surface load, reduction of salt bridges, and changes in the active site, all characteristics that facilitate functional activity in low pH.

Acidophilic microbes are widely used in biotechnology, especially in biomining, where acidophiles are used in the extraction and recovery of elements such as zinc, copper, and nickel, as well as precious metals and uranium, which may be present in small quantities in the environment or in mining wastewater [7, 92, 93]. Techniques using acidophiles have the advantages of low investment and operational costs, saving inputs such as the acids and oxidizing agents used in conventional hydrometallurgical processes, since the microorganism itself produces these byproducts, and low energy requirements, besides being more sustainable because no sulfur dioxide is emitted [93]. For these biotechnologies, the most widely used species are members of the extremophile genera *Acidithiobacillus*, *Acidiphilium*, *Acidiferrobacter*, *Ferrovum*, *Ferroplasma*, *Alicyclobacillus*, *Sulfobacillus*, *Sulfolobus*, *Acidianus*, and *Metallosphaera* [94, 95].

3.2.4 Alkaliphiles

Microorganisms that require pH levels above 9 are called alkaliphiles. Other microorganisms, called alkalitolerant, have optimal growth below pH 9 but tolerate and grow well at high pH (>9) [27]. Some alkaliphiles can grow even at pH 13, although microorganisms able to tolerate such high pH levels are uncommon [96, 97].

Alkaline as well as acidic environments may be of natural or anthropic origin, such as soda lakes, insect guts and pancreatic juice, gardens, farm soils, manure, underground alkaline water, and alkaline residues from the cement, paper and pulp industries, as well as in the fields of electroplating, textile processing and food processing [3, 98, 99].

The genera isolated from these environments belong to all domains of life, but mainly to the archaean genera *Natronomonas*, *Natrialba*, *Natronolimnobius*, and *Halorubrum*; bacteria of the genera *Bacillus*, *Pseudomonas*, *Rhodobacterium*, and *Methylobacter*; the cyanobacteria *Cyanospira*, *Chroococcus*, *Synechococcus*, and *Spirulina*; and the actinomycete *Bogoriella* [65].

Several investigators have studied how alkaliphiles survive these extreme environments and what adaptive strategies they employ [84, 100]. Most studies have focused on the mechanisms of cytoplasmic pH reduction, analysis of cell-wall components, and the bioenergetic challenges faced by the membrane [101]. Krulwich et al. [102] observed that cell-wall polymers such as teichuronopeptide play an important role in supporting pH homeostasis, as well as the S layer present in some microorganisms, which confers an advantage in high pH levels. An increase in the regulation of the enzyme deaminase, which produces cytoplasmic acids, was observed in both physiological and transcriptome and proteome analyses [6, 103].

Alkaliphiles also have the ability to produce enzymes that remain active and stable under alkaline conditions. According to Zhao et al. [104], the adaptive mechanisms of alkaliphilic enzymes are still poorly understood, although it is known that the primary structure of the proteins contains more hydrophobic residues.

The industrial applications of these extremophiles are based mainly on their enzymes, which are stable under high pH conditions [105]. Industrial applications of this bioproduct are primarily for detergent production, account for about 30% of the total global production of enzymes, and have been in use for more than 50 years [106].

According to Mamo and Mattiason [106], alkaline detergents are the most commonly used for cleaning dishes and laundry, as alkalinity increases dirt removal and washing performance; to remove protein stains, proteases are used. A large number of alkaline-active proteases have been studied to date, and most of the commercially available detergent proteases are derived from *Bacillus* spp.

Another application of alkaline enzymes is in the textile industry, where cellulases adapted to remain active at high pH are used in cotton-based textile processing. These give a better textile finish, facilitate removal of the dye in the washing process, and provide a faster and lower-cost process with low environmental impact [107]. Proteases and alkaline cellulases are used in the food industry, wine and beer fermentation, and for processing beverages such as orange juice [108].

3.2.5 Piezophiles

Piezophiles are organisms that colonize the deep biosphere, such as ocean depths (major oceans, except the Arctic Ocean), hydrothermal crevices, and the lithosphere, all high-pressure environments. Piezophiles can be classified according to their preference [109, 110]. Piezo-tolerant organisms tolerate pressures higher than atmospheric pressure, i.e. ranging from 0.1 to 10 MPa; piezophiles require 10–50 MPa for optimal growth; and hyperpiezophiles require pressures above 50 MPa [109, 111].

Organisms that develop in this extreme environment are generally anaerobic or facultative, gram-negative, and psychrophilic bacteria belonging to genera of Gammaproteobacteria, such as *Photobacterium*, *Shewanella*, *Colwellia*, *Psychromonas*, *Thioprofundum*, and *Moritella*, as well as piezophilic archaea, which are mainly hyperthermophiles of the kingdoms Euryarchaeota and Crenarchaeota, of which the order Thermococcales is the most important [112, 113].

The biochemistry and physiology of these extreme organisms are largely unknown. However, using high-pressure culture technologies combined with omics tools and DNA recombination methods, scientists are beginning to understand the adaptive strategies of these organisms along with their ecological functions and evolutionary processes [8].

The effects of temperature on various biological systems are widely known, but the effects of pressure have not been thoroughly analyzed. It is known that high pressure profoundly affects the cell structure and its activity, which can lead to cell death [13]. Biochemical and genomic studies give only a fragmentary view of the adaptive mechanisms in piezophiles and do not fully reveal how microbial adaptation occurs at high pressures. Adaptation may be related to the modification of certain genes, a profound reorganization of the genome or adjustment of gene expression, metabolic pathways, or a physiological response to stress. It is suggested that, contrary to what occurs in thermophilic microorganisms, adaptations to high pressure are scattered within the genome and may occur only in a small fraction of the genes [8].

In order to neutralize the effects of high pressure on the cells, the piezophiles alter the membrane composition, especially the lipid bilayer, with increased amounts of monounsaturated fatty acids in the lipids (70%), and increased fluidity of the membrane. Only piezophilic bacteria can synthesize long-chain polyunsaturated fatty acids, either eicosapentaenoic acid, docosahexaenoic acid, or both, neither of which is found in surface-dwelling bacteria [109].

Studies by Tamegai et al. [114] and Amrani et al. [115] showed that piezophilic adaptation involves major cellular processes, such as the transport of solutes including amino acids, amino-acid metabolism, and the pathways involved in energy production. Michoud and Jebbar [111] observed by means of molecular analysis that the responses of microorganisms to high and low pressures are very similar, except for certain functions such as transmembrane ATPases and replication. High pressure also affects the pathways of chemotaxis, translation via ribosomes, CRISPR-Cas, and energy metabolism, with regulation of hydrogenases, oxidoreductases, and formation metabolism (anaerobes).

The biotechnological applications of piezophiles have not yet been extensively investigated, and only their potential for biotechnological use has been suggested. According to Abe and Horikoshi [116], because of their adaptation to extreme-pressure environments, piezophilic and piezotolerant microorganisms have considerable potential for biotechnological uses in the pharmaceutical and medical industries, particularly in the area of antibiotics. Abe and Horikoshi [116] believed that advances in molecular biology will lead to the discovery of a variety of industrial applications of piezophiles, which are economical and environmentally clean processes, such as for pasteurization of food, in the pharmaceutical industry with the inactivation of pathogens, and the use of DNA polymerases from deep-sea hydrothermal species [117].

3.2.6 Radioresistants

The Earth's biosphere is constantly bombarded by radiation of natural or anthropic origin. Radiation is energy in the form of particles or electromagnetic waves, including gamma rays, X-rays, ultraviolet (UV) radiation, and radio waves [12, 118]. Most radiation is absorbed by the Earth's atmosphere, including the ozone layer, before it reaches the surface; however, in some regions, this layer is reduced, such as near Poland, Australia, New Zealand, and areas exposed to substantial amounts of sunlight and more intense radiation, such as Ecuador and Antarctica [12].

According to Pavlopoulou et al. [119], the most notable category among extremophiles is the radiation-resistant organisms, also called radioresistant. To date, no microorganisms that can be classified as true radiophiles, i.e. that require intense radiation to develop, have been described; however, radioresistant microorganisms can survive extremely high doses of radiation [120].

The most serious effect of radiation is severe oxidative damage to macromolecules, including proteins, lipids, and nucleic acids, generated by oxygen-reactive and nitrogen-reactive species [118, 121]. Microorganisms can resist ionization and other forms of radiation, or at least deal with their damage and resulting mutations. Such adaptations were developed by radioresistant bacteria and archaea and are related to strategies to prevent direct or indirect radiation damage to these molecules, mainly proteins and DNA [12].

Among these strategies are polyploidy; the induction of specific genes, proteins, and enzymes for DNA repair as well as cell protection; the presence of free antioxidant compounds in the cytoplasm; the presence of pigments (carotenoids and melanin); and an increased iron/manganese ratio in proteins [122–124]. Many members of the phylum Firmicutes are able to form spores, which, in addition to protecting against DNA damage, contain acid-soluble proteins and have active repair mechanisms to enable good germination [125].

The bacterial family Deinococcaceae, particularly the genus *Deinococcus*, is the best-studied group, and consequently, more information about its radiation-resistant characteristics is available [126, 127]. Other genera of this family are also able to withstand high levels of radiation, such as *Kineococcus*, *Chroococcidiopsis*, *Pseudomonas*, *Hymenobacter*, *Sphingomonas*, *Flavobacterium*, *Thermococcus*, *Pyrococcus*, and *Janthinobacterium* [128]. Marizcurrena et al. [12], searching for radioresistant bacteria in Antarctica, suggested that this extreme continent has a very diverse group of bacteria with resistance to UV wavelengths. Turick et al. [129] stated that melanized microorganisms dominate environments with high levels of ionizing radiation, such as uranium-contaminated soils, since they produce photoprotective pigments and absorb radiation.

The biotechnological potential of radioresistant microorganisms is being evaluated. According to Gabani and Singh [130], the use of these extremophiles in the bioenergy area is not extensively studied, mainly because researchers have not found an inexpensive and efficient process to degrade carbohydrates for bioethanol production. Studies on the application of radioresistants to bioremediation of nuclear waste include those of Kim et al. [131], who observed that the Proteobacteria *Geobacter* sp. and *Rhodoferax ferrireducens* have a potential metabolism to reduce radioisotopes, and of Seyrig [132], who reported that marine algae of the genus *Cystoseira* effectively adsorbed uranium radioisotopes.

The widely studied radioresistant *Deinococcus radiodurans* was effective in the bioremediation of heavy metals and was able to remove 70% of a 1-mM uranium solution [133]. This species is a source of new enzymes and has shown proteolytic activity on gelatin, casein, and keratin and also produces carotenoids that can be used in the pharmaceutical and cosmetics sectors [134, 135].

3.3 Approaches and new trends

Although microorganisms originating from extreme environments have attractive characteristics for biotechnological application, it is difficult to find biofilms of microbial origin that produce high yields when cultured on a large scale for industrial applications. Part of this is due to the difficulty of mimicking extreme conditions in the laboratory and allowing these microorganisms to synthesize the target bioproducts. This attempt requires a large investment of funds, energy, development, and adaptation of the required culture methods, which makes the process more expensive and not widely marketed due to the excessive cost [136, 137].

In order to overcome these problems, efforts to optimize the production of bioproducts such as organic molecules and enzymes have focused on two aspects: (1) to improve the catalytic activity of existing bioproducts and/or (2) to find new bioproducts.

One of the strategies that have been studied to improve the use of extremophilic enzymes in industry is protein engineering. A major research strategy that has been developed in the last two decades aims to increase the efficiency of enzymes for industrial purposes by optimizing the amino acid sequence, making use of molecular methods to produce stable enzymes to meet specific industrial requirements [20, 138].

Protein engineering has become a powerful means to alter or improve enzyme catalysis and is currently among the successful technological approaches in biotechnology, able to generate valuable intellectual property. This engineering uses two methods: (1) directed evolution, which is a random mutagenesis applied to a protein and (2) rational design of proteins, which makes it necessary to determine the structure and function of a protein in order to change its characteristics [47]. These methods have been applied successfully to increase protein selectivity and thermostatic activity [139].

In protein engineering, mutation is the key to exploit protein function, generating mutations by physical and chemical methods such as the use of UV irradiation, gamma rays, and nitrous acid at particular sites to achieve new functions or design a completely new product that can yield spectacular results for better integration into a process [138]. With the advances in mutagenesis, specifically in the area of enzyme modification, structures can be altered to achieve the desired effect for particular applications.

The main application of mutagenesis is to introduce new properties, such as improved specificity, stability, solubility, and expression for the desired biocatalyst [140]. Shu et al. [141] irradiated psychrophilic bacteria to produce a mutated strain that increased lipase yield by up to 10 times. Xu et al. [142] improved the molecular properties and activity of the enzyme α-galactosidase of the deep-sea *Bacillus megaterium*. Yasukawa and Inouye [143] succeeded in increasing the stability and thermolysin catalysis 5- to 10-fold compared to the non-mutated enzyme, and Chi et al. [144] achieved an amylase that is resistant to chemical oxidation.

Lidan et al. [145], through mutagenesis, obtained an increase in lactic acid production at pH 3.8 by *Lactobacillus pentosus* ATCC 8041, while the non-mutated strain did not grow at this low pH. Zhang et al. [146] obtained positive results using mutagenesis in *Candida antarctica*, i.e. a simultaneous increase in the activity and stability of a psychrophilic lipase. Similarly, Liu et al. [147] used mutagenesis to adapt an alkaline protease for production at a psychrophilic temperature.

Another strategy is to clone and express thermophilic and psychrophilic enzymes in mesophilic hosts, such as *Escherichia coli* and *Saccharomyces cerevisiae*, that can be used to improve the enzymes, increasing production on an industrial scale, reducing the energy expenditure to maintain the required temperature in the process, and without losing their functional characteristics [148].

Some successes include that of Chakraborty et al. [149], who cloned the thermophilic esterase of *Archaeoglobus fulgidus* overexpressed in *E. coli*, and of Hayashi [150], who cloned the gene coding for thermostable α-amylase from the extremophilic archaeon *Thermococcus hydrothermalis* and inserted it in *E. coli*. Also, several amylopullulanase enzymes were isolated from thermophilic microorganisms and cloned in *E. coli* and *Bacillus subtilis*, for expression at mesophilic temperatures [151].

With respect to the psychroenzymes expressed in vectors, Alias et al. [152] cloned and expressed in *Pichia pastoris* a new serine protease isolated from *Glaciozyma antarctica* PI12, which shows high activity at 20°C. Ferrer et al. [153] expressed an esterase and two chaperones from the psychrophilic bacterium *Oleispira antarctica* in *E. coli* and found 180-fold higher expression by the recombinant strains.

As society's demands for products from extremophiles grow, researchers are applying different approaches in the search for new bioproducts, such as the study of "omics" – metagenomics, metatranscriptomics, and metaproteomics. This search is made possible only through technological advances, the evolution of tools for data processing and analysis, and the availability of DNA sequence information in public databases [20]. These technologies allow rapid identification of new metabolites that can also be used for the generation of new synthetic products and knowledge of metabolic pathways with optimal performance, an increasingly important and widely used approach [154].

Complete sequencing of genomes from extremophiles has made it possible to analyze a substantial number of genes coding for bioproducts, which are of interest for basic and applied research [155, 156]. The same is true for metagenomic studies of extreme environments, which have contributed substantially to the understanding of strategies used by extremophiles and have been used to search and mine for microbial products [157].

Metagenomics using shotgun sequencing allows the entire genetic material in a sample to be analyzed, omitting the cloning steps and resulting in sequence after sequence, thus allowing the prediction of genes and the description and functional assignments for the metagenomic sequences [158]. Additionally, knowledge of the metagenome allows a reconstruction of the genomes present in an environmental sample and a functional analysis of the community members [17].

Because of the impact of the market for enzymes and its generation of billions of dollars annually, and the efforts made toward sustainable development by companies targeting products of biological origin, these new tools of molecular biology and bioinformatics have been used in the search for functional genes from microorganisms of interest [43, 159, 160].

Several industrially important extreme bioproducts have been discovered using metagenomics. With the advent of this powerful tool, several research groups have been exploring these data, searching for genes that encode bioproducts for industrial application, especially enzymes [161–163]. For example, Kennedy et al. [164] listed a series of extremophilic enzymes from marine environments, including lipases, esterases, proteases, amylases, cellulases, DNA polymerases, proteases, and phytases, among others. All of these were discovered through functional metagenomics, by analyzing genes with specific functions.

Using the same process, Ferrandi et al. [165] discovered two new epoxide hydrolases in thermophilic soil samples with the aid of bioinformatics, which identified the genes coding for these enzymes, which are of pharmaceutical interest and were cloned and overexpressed in *E. coli*. Lopez-Lopez et al. [166] mined, in thermophilic metagenomes, genes coding for lipase and esterase enzymes.

Several hundred commercially useful enzymes (CUEs), all of which have been identified by metagenomic analyses since 2010, have been made public and can be accessed through the MetaBioMe database. These CUEs are categorized in nine application categories, including agriculture, energy, food and nutrition, biotechnology, biosensory, environment, and health [167].

Other advances in metagenomic methodology are the use of more suitable expression vectors, host strains, selection and selection methods, with faster sequencing, computational modeling, and bioinformatics. Together, these techniques will help to discover and determine the functions of new genes for biotechnological applications [168].

Proteomics has also emerged as a new tool for functional genomics analyses. Proteomic studies of extremophiles can provide information on expressed proteins, novel proteins, and the number of proteins induced under specific conditions [169], as well as elucidate the physiological characteristics of extremophiles and categorize proteins according to their roles as biological agents, with the help of bioinformatics [170]. Challenges remain, particularly in applying proteomics to extremophiles, due to the difficulty of culturing them; however, new methods and techniques have recently been developed to partly overcome these limitations [169, 171].

As mentioned in the other subtopics of this chapter, extremophiles have unique metabolic activities and survival mechanisms and constitute a valuable resource for the discovery of new enzymes and biological compounds, especially those adapted to functioning under extreme conditions. Their enzymes, such as psychrophilic lipases, alkaliphilic cellulases, and thermostatic DNA polymerases, have broad applications in the biotechnology industry. Therefore, improvement of screening methods using proteomics is essential for the identification of new proteins in extremophiles [170, 172].

Another area that has attracted increasing interest is synthetic biology, which, in several respects, is related to genetic engineering but also integrates knowledge in the areas of microbiology, systems biology, and bioinformatics. This field aims to design and build new functions and biological systems that are not found in nature or to improve functions of interest through the creation of synthetic genomes. Synthetic biology has been a valuable tool for the optimization of processes related to products from microbes living in extreme environments [173, 174].

3.4 Conclusions

Microorganisms that live in the extreme habitats in the biosphere have great biotechnological potential because of their unique characteristics that allow them to be active in such environmental conditions. These characteristics are finding wide use in various industrial sectors that use processes that require extreme conditions.

Access to the bioproducts of extremophilic microorganisms continues to depend on the ability to culture them. However, in the post-genomic era, this process could be optimized, as could the process for developing new alternatives in the search for these products.

Much remains to be done in the field of genetic engineering, to increase the accessibility of biotechnology tools for use in molecular biology, structural biology, and bioinformatics.

There is a growing need to improve the tools that optimize production on industrial scale of the bioproducts of extremophilic microorganisms, and great advances should be achieved with recent developments in the areas of systems biology and synthetic biology. These approaches should help in the search for genes, metabolic pathways and information, without the need to culture microorganisms, since culturing requires the arduous and almost impossible task of trying to mimic extreme conditions in a petri dish.

References

[1] Horikoshi K, Bull AT. Prologue: definition, categories, distribution, origin and evolution, pioneering studies, and emerging fields of extremophiles. In: Horikoshi K, Antranikaian G, Bull AT, Robb FT, Stetter KO, editors. Extremophiles handbook. Tokyo, Japan: Springer; 2011. p. 4–15.

[2] Rothschild LJ, Mancinelli RL. Life in extreme environments. Nature. 2001;409:1092–101.

[3] Horikoshi K. Enzymes isolated from alkaliphiles. In: Horikoshi K, Antranikaian G, Bull AT, Robb FT, Stetter KO, editors. Extremophiles handbook. Tokyo, Japan: Springer; 2011. p. 164–77.

[4] Stan-Latter H. Physico-chemical boundaries of life. In: Stan-Latter H, Fendrihan S, editors. Adaptation of microbial life to environmental extremes. New York, NY: Springer-Verlag; 2012. p. 1–14.

[5] De Maayer P, Anderson D, Cary C, Cowan DA. Some like it cold: understanding the survival strategies of psychrophiles. EMBO Rep. 2014;15:508–17.

[6] Padan E, Bibi E, Ito M, Krulwich TA. Alkaline pH homeostasis in bacteria: New insights. Biochim Biophys Acta. 2005;1717:67–88.

[7] Johnson DB, Schippers A. Editorial: recent advances in acidophile microbiology: fundamentals and applications. Front Microbiol. 2017;8:428.

[8] Abe F. Effects of high hydrostatic pressure on microbial cell membranes: structural and functional perspectives. Subcell Biochem. 2015;72:371–81.

[9] Enache M, Popescu G, Itoh T, Kamekura M. Halophilic microorganisms from man-made and natural hypersaline environments: physiology, ecology, and biotechnological potential. In: Stan-Latter H, Fendrihan S, editors. Adaptation of microbial life to environmental extremes. New York, NY: Springer-Verlag; 2012. p. 173–97.

[10] Leong SL, Pettersson OV, Rice T, Hocking AD, Schnürer J. The extreme xerophilic mould *Xeromyces bisporus* – growth and competition at various water activities. Int J Food Microbiol. 2011;145(1):57–63.

[11] Pitt JI, Hocking AD. Fungi and food spoilage. New York, NY: Springer Science+Business Media; 2009.

[12] Marizcurrena JJ, Morel MA, Braña V, et al. Searching for novel photolyases in UVC-resistant Antarctic bacteria. Extremophiles. 2017;21(2):409–18.

[13] Capece MC, Clark E, Saleh JK, Halford D, Heinl N, Hoskins S, Rothschild LJ. Polyextremophiles and the constraints for terrestrial habitability. In: Seckbach J, Oren A, Stan-Latter H, editors. Polyextremophiles: life under multiple forms of stress. Dordrecht, the Netherlands: Springer; 2013. p. 3–60.

[14] Straub CT, Zeldes BM, Schut, GJ, Adams MWW, Kelly RM. Extremely thermophilic energy metabolisms: biotechnological prospects. Curr Opin Biotechnol. 2017;45:104–12.

[15] Sing, OV. Extremophiles: sustainable resources and biotechnological implications. Hoboken, NJ: Wiley-Blackwell; 2012.

[16] Sarmiento F, Peralta R, Blamey JM. Cold and hot extremozymes: industrial relevance and current trends. Front Bioeng Biotechnol. 2015;3:148.

[17] Baweja M, Nain L, Kawarabayasi Y, Shukla P. Current Technological improvements in enzymes toward their biotechnological applications. Front Microbiol. 2016;7:965.

[18] Rampelotto PH. Biotechnology of extremophiles: advances and challenges (grand challenges in biology and biotechnology). New York, NY: Springer; 2016.

[19] Suriya J, Bharathiraja S, Krishnan M, Manivasagan P, Kim SK. Extremozymes from marine actinobacteria. Adv Food Nutr Res. 2016;79:43–66.

[20] Khan M, Sathya TA. Extremozymes from metagenome: potential applications in food processing. Crit Rev Food Sci Nutr. 2017;12:1–9.

[21] Madigan MT, Martinko JM, Dunlap PV, Clarck DP. Brock biology of microorganisms. San Francisco, CA: Benjamin Cummings; 2014.

[22] Takai K, Nakamura K, Toki T, et al. Cell proliferation at 122 degrees C and isotopically heavy CH_4 production by a hyperthermophilic methanogen under high-pressure cultivation. Proc Natl Acad Sci USA. 2008;105:10949–54.

[23] Stetter KO. A brief history of the discovery of hyperthermophilic life. Biochem Soc Trans. 2013;41(1):416–20.

[24] Herbold CW, McDonald IR, Cary C. Microbial ecology of geothermal habitats in Antarctica. In: Cowan DA, editor. Antarctic Terrestrial Microbiology: physical and biological properties of Antarctic soils. New York, NY: Springer; 2014. p. 181–215.

[25] Singh B, Poças-Fonseca MJ, Johri BN, Satyanarayana T. Thermophilic molds: Biology and applications. Crit Rev Microbiol. 2016;42(6):985–1006.

[26] Yand DC, Blair KM, Salamaa NR. Staying in shape: the impact of cell shape on bacterial survival in diverse environments. Microbiol Mol Biol Rev. 2016;80:1.

[27] Moriya T, Hikota T, Yumoto I, Ito T, Terui Y, Yamagishi A, Oshima T. *Calditerricola satsumensis* gen. nov., sp. nov. and *Calditerricola yamamurae* sp. nov., extreme thermophiles isolated from a high-temperature compost. Int J Syst Evol Microbiol. 2011;61:631–6.

[28] Lopez-Vazquez CM, Kubare M, Saroj DP, Chikamba C, Schwarz J, Daims H, Brdjanovic D. Thermophilic biological nitrogen removal in industrial wastewater treatment. Appl Microbiol Biotechnol. 2014;98(2):945–56.

[29] Satyanarayana T, Littlechild J, Kawarabayasi Y. Thermophilic microbes in environmental and industrial biotechnology: biotechnology of thermophiles. New York, NY: Springer; 2013.

[30] Ranawat P, Rawat S. Radiation resistance in thermophiles: mechanisms and applications. World J Microbiol Biotechnol. 2017;33:112.

[31] Koga, Y. Thermal adaptation of the archaeal and bacterial lipid membranes. Archaea. 2012;789652:6.

[32] Siliakus MF, van der Oost J, Kengen SWM. Adaptations of archaeal and bacterial membranes to variations in temperature, pH and pressure. Extremophiles. 2017;21(4):651–70.

[33] Mandelli F, Miranda VS, Rodrigues E, Mercadante AZ. Identification of carotenoids with high antioxidant capacity produced by extremophile microorganisms. World J Microbiol Biotechnol. 2012;28:1781–90.

[34] Mandelli F, Yamashita F, Pereira JL, Mercadante AZ. Evaluation of biomass production, carotenoid level and antioxidant capacity produced by *Thermus filiformis* using fractional factorial design. Braz J Microbiol. 2012;43:1.

[35] Hara M, Yamano Y, Sakai Y, Kodama E, Hoshino T, Ito M, Miyake J. Stabilization of liposomal membranes by carotenoids: zeaxanthin, zeaxanthin glucoside and thermozeaxanthin. Mater Sci Eng C. 2008;28(2):274–9.

[36] Aung HL, Samaranayaka CU, Enright R, Beggs KT, Monk BC. Characterisation of the DNA gyrase from the thermophilic eubacterium *Thermus thermophilus*. Protein Expr Purif. 2015;107:62–7.

[37] van Wolferen M, Ajon M, Driessen AJ, Albers SV. How hyperthermophiles adapt to change their lives: DNA exchange in extreme conditions. Extremophiles. 2013;17(4):545–63.

[38] Wu H, Zhang Z, Hu S, Yu J. On the molecular mechanism of GC content variation among eubacterial genomes. Biol Direct. 2012;7:2.

[39] Dong H, Mukaiyama A, Tadokoro T, Koga Y, Takano K, Kanaya S. Hydrophobic effect on the stability and folding of a hyperthermophilic protein. J Mol Biol. 2008;378(1):264–72.

[40] Cava F, Hidalgo A, Berenguer J. *Thermus thermophilus* as biological model. Extremophiles. 2009;13:213–31.

[41] Large AT, Goldberg MD, Lund PA. Chaperones and protein folding in the archaea. Biochem Soc Trans. 2009;37(1):46–51.

[42] Vieille C, Zeikus JG. Hyperthermophilic enzymea: sources, uses and molecular mechanisms for thermostability. Microb Mol Biol Rev. 2001;65(1):1–43.

[43] Siddiqui KS. Some like it hot, some like it cold: temperature dependent biotechnological applications and improvements in extremophilic enzymes. Biotechnol Adv. 2015;33(8):1912–22.

[44] Kongpol A, Kato J, Vangnai AS. Isolation and characterization of *Deinococcus geothermalis* T27, a slightly thermophilic and organic solvent-tolerant bacterium able to survive in the presence of high concentrations of ethyl acetate. FEMS Microb Lett. 2008;286:227–35.

[45] Kongpol A, Pongtharangkul T, Kato J, Honda K, Ohtake H, Vangnai AS. Characterization of an organic-solvent-tolerant *Brevibacillus agri* strain13 able to stabilize solvent/water emulsion. FEMS Microbiol Lett. 2009;297:225–33.

[46] Mehta R, Singhal P, Singh H, Damle D, Sharma, AK. Insight into thermophiles and their wide-spectrum applications. 3 Biotech. 2016;6(1):81.

[47] Bornscheuer UT, Huisman GW, Kazlauskas RJ, Lutz S, Moore JC, Robins K. Engineering the third wave of biocatalysis. Nature. 2012;485:185–94.

[48] Khan FI, Lan D, Durrani R, Huan W, Zhao Z, Wang Y. The lid domain in lipases: Structural and functional determinant of enzymatic properties. Front Bioeng Biotechnol. 2017;5:16.

[49] Miyazaki K. A hyperthermophilic laccase from *Thermus thermophilus* HB27. Extremophiles. 2005;9(6):415–425.

[50] Mandelli F, Franco-Cairo JPL, Citadini APS, et al. The characterization of a thermostable and cambialistic superoxide dismutase from *Thermus filiformis*. Lett Appl Microbiol. 2013;57(1):40–6.

[51] Simair AA, Qureshi AS, Khushk I, et al. Production and partial characterization of α-amylase enzyme from *Bacillus sp*. bcc 01-50 and potential applications. Biomed Res Int. 2017;9173040.

[52] Sayer C, Szabo Z, Isupov MN, Ingham C, Littlechild JA. The structure of a novel thermophilic esterase from the planctomycetes species, *Thermogutta terrifontis* reveals an open active site due to a minimal "cap" domain. Front Microbiol. 2015;6:1294.

[53] Panda MK, Sahu MK, Tayung K. Isolation and characterization of a thermophilic *Bacillus* sp. with protease activity isolated from hot spring of Tarabalo, Odisha, India. Iran J Microbiol. 2013;5(2):159–65.

[54] Potprommanee L, Wang XQ, Han YJ, et al. Characterization of a thermophilic cellulase from *Geobacillus* sp. HTA426, an efficient cellulase-producer on alkali pretreated of lignocellulosic biomass. PLoS ONE. 2017;12(4):e0175004.

[55] Chakdar H, Kumar M, Pandiyan K, Singh A, Nanjappan K, Kashyap PL, Srivastava AK. Bacterial xylanases: biology to biotechnology. 3 Biotech. 2016;6(2):150.

[56] Littlechild JA. Enzymes from extreme environments and their industrial applications. Front Bioeng Biotechnol. 2015;3:161.

[57] Gomes E, Souza AR, Orjuela GL, et al. Applications and benefits of thermophilic microorganisms and their enzymes for industrial biotechnology. In: Schmoll M, Dattenböck C, editors. Gene expression systems in fungi: advancements and applications. New York, NY: Springer; 2016. p. 459–92.

[58] Egorova K, Antranikian G. Industrial relevance of thermophilic Archaea. Curr Opin Microbiol. 2005;8(6):649–55.

[59] Cornec L, Robineau J, Rolland JL, Dietrich J, Barbier G. Thermostable esterases screened on hyperthermophilic archaeal and bacterial strains isolated from deep-sea hydrothermal vents: characterization of esterase activity of a hyperthermophilic archaeum, *Pyrococcus abyssi*. J Mar Biotechnol. 1998;6:104–10.

[60] Seong-Ae Y, Ryu SI, Lee SB, Moon TW. Purification and characterization of branching specificity of a novel extracellular amylolytic enzyme from marine hyperthermophilic *Rhodothermus marinus*. J Microbiol Biotechnol. 2008;18:457–64.

[61] Lundberg KS, Shoemaker DD, Adams MWW, et al. High-fidelity amplification using a thermostable DNA polymerase isolated from *Pyrococcus furiosus*. Gene. 1991;108:1–6.
[62] Feller G. Psychrophilic enzymes: From folding to function and biotechnology. Scientifica. 2013;512840.
[63] Rodrigues DF, Tiedje JM. Coping with our cold planet. Appl Environ Microbiol. 2008;74(6):1677–86.
[64] Margesin R, Miteva V. Diversity and ecology of psychrophilic microorganisms. Res Microbiol. 2011;162:346–61.
[65] Canganella F, Wiegel J. Anaerobic thermophiles. Life. 2014;4:77–104.
[66] Struvay C, Feller G. Optimization to low temperature activity in psychrophilic enzymes. Int J Mol Sci. 2012;13(9):11643–65.
[67] Casanueva A, Tuffin M, Cary C, Cowan DA. Molecular adaptations to psychrophily: the impact of "omic" technologies. Trends Microbiol. 2010;18:374–81.
[68] Christner BC. Bioprospecting for microbial products that affect ice crystal formation and growth. Appl Microbiol Biotechnol. 2010;85(3):481–9.
[69] Chattopadhyay MK. Mechanisms of bacterial adaptation to low temperatures. J Biosci. 2006;31:157–65.
[70] Metpally RP, Reddy BV. Comparative proteome analysis of psychrophilic versus mesophilic bacterial species: Insights into the molecular basis of cold adaptation of proteins. BMC Genomics. 2009;10:11.
[71] Chen Z, Yu H, Li L, Hu S, Dong X. The genome and transcriptome of a newly described psychrophilic archaeon, *Methanolobus psychrophilus* R15, reveal its cold adaptive characteristics. Environ Microbiol Rep. 2012;4(6):633–41.
[72] Ayala-del-Río HL, Chain PS, Grzymski JJ, et al. The genome sequence of *Psychrobacter arcticus* 273-4, a psychroactive Siberian permafrost bacterium, reveals mechanisms for adaptation to low-temperature growth. Appl Environ Microbiol. 2010;76(7):2304–12.
[73] Cavicchioli R, Charlton T, Ertan H, Mohd Omar S, Siddiqui KS, Williams TJ. Biotechnological uses of enzymes from psychrophiles. Microb Biotechnol. 2011;4(4):449–60.
[74] Stavros P, Malecki PH, Theodoridou M, Rypniewski W, Vorgias CE, Nounesis G. The stability of the TIM-barrel domain of a psychrophilic chitinase. Biochem Biophys Rep. 2015;3:108–116.
[75] Sharma D, Sharma B, Shukla AK. Biotechnological approach of microbial lipase: a review. Biotechnology. 2011;10:23–40.
[76] Siddiqui KS, Cavicchioli R. Cold-adapted enzymes. Annu Rev Biochem. 2006;75:403–33.
[77] Ueda M, Goto T, Nakazawa M, Miyatake K, Sakaguchi M, Inouye K. A novel cold-adapted cellulase complex from *Eisenia foetida*: characterization of a multienzyme complex with carboxymethylcellulase, β-glucosidase, β-1,3 glucanase, and β-xylosidase. Comp Biochem Physiol B Biochem Mol Biol. 2010;157:26–32.
[78] Wang F, Hao J, Yang C, Sun M. Cloning, expression, and identification of a novel extracellular cold-adapted alkaline protease gene of the marine bacterium strain YS-80-122. Appl Biochem Biotechnol. 2010;162:1497–505.
[79] Fendriham S, Negoiţă TG. Psychrophilic microorganisms as important source for biotechnological processes. In: Stan-Latter H, Fendrihan S, editors. Adaptation of microbial life to environmental extremes. New York, NY: Springer-Verlag; 2012. p. 133–72.
[80] Aguilera A. Eukaryotic organisms in extreme acidic environments, the Río Tinto Case. Life. 2013;3(3):363–74.
[81] Johnson DB, Hallberg KB. The microbiology of acidic mine waters. Res Microbiol. 2003; 154(7):466–73.
[82] Bertoldo C, Dock C, Antranikian G. Thermoacidophilic microorganisms and their novel biocatalysts. Eng Life Sci. 2004;4:521–31.
[83] Oren A. Industrial and environmental applications of halophilic microorganisms. Environ Technol. 2010;31:825–34.

[84] Sharma A, Parashar D, Satyanarayan T. Acidophilic microbes: biology and applications. In: Rampelotto PH, editors. Biotechnology of extremophiles: advances and challenges (grand challenges in biology and biotechnology). New York, NY: Springer; 2016. p. 215–42.
[85] Baker-Austin C, Dopson M. Life in acid: pH homeostasis in acidophiles. Trends Microbiol. 2007;15(4):165–71.
[86] Macalady JL, Vestling MM, Baumler D, Boekelheide N, Kaspar CW, Banfield JF. Tetraether-linked membrane monolayers in *Ferroplasma* spp: a key to survival in acid. Extremophiles. 2004;8:411–9.
[87] Mangold S, Valdés J, Holmes D, Dopson M. Sulfur metabolism in the extreme acidophile *Acidithiobacillus caldus*. Front Microbiol. 2011;2:17.
[88] Spiekermann P, Rehm BHA, Kalscheuer R, Baumeister D, Steinbüchel A. A sensitive, viable-colony staining method using Nile red for direct screening of bacteria that accumulate polyhydroxyalkanoic acids and other lipid storage compounds. Arch Microb. 1999;171:73–80.
[89] Reed CJ, Lewis H, Trejo E, Winston V, Evilia C. Protein adaptations in archaeal extremophiles. Archaea. 2013;373275.
[90] Huang Y, Krauss G, Cottaz S, Driguez H, Lipps G. A highly acid-stable and thermostable endo-b-glucanase from the thermoacidophilic archaeon *Sulfolobus solfataricus*. Biochem J. 2005;385:581–588.
[91] Michaux C, Pouyez J, Mayard A, Vandurm P, Housen I, Wouters J. Structural insights into the acidophilic pH adaptation of a novel endo-1,4-β-xylanase from *Scytalidium acidophilum*. Biochimie. 2010;92(10):1407–15.
[92] Harrison STL. Biotechnologies that utilize acidophiles. In: Quatrini R, Johnson DB, editors. Acidophiles. Life in extremely acidic environments. Norfolk, UK: Caister Academic Press; 2016. p. 265–83.
[93] Jerez CA. Biomining of metals: how to access and exploit natural resource sustainably. Microbial Biotechnol. 2017;10(5);1191–3.
[94] Hedrich S, Schlömann M, Johnson DB. The iron-oxidizing proteobacteria. Microbiology. 2011;157:1551–64.
[95] Vera M, Krok B, Bellenberg S, Sand W, Poetsch A. Shotgun proteomics study of early biofilm formation process of *Acidithiobacillus ferrooxidans* ATCC 23270 on pyrite. Proteomics. 2013;13:1133–44.
[96] Grum-Grzhimaylo AA, Georgieva ML, Bondarenko SA, Debets AJM, Bilanenko EN. On the diversity of fungi from soda soils. Fungal Divers. 2016;76(1):27–74.
[97] Dhakar K, Pandey A. Wide pH range tolerance in extremophiles: towards understanding an important phenomenon for future biotechnology. Appl Microbiol Biotechnol. 2016;100(6):2499–510.
[98] Yumoto I, Hishinuma-Narisawa M, Hirota K, Shingyo T, Takebe F, Nodasaka Y, Matsuyama H, Hara I. *Exiguobacterium oxidotolerans* sp. nov., a novel alkaliphile exhibiting high catalase activity. Int J Syst Evol Microbiol. 2004;54:2013–7.
[99] Roadcap GS, Sanford RA, Jin QS, Pardinas JR, Bethke CM. Extremely alkaline (pH > 12) ground water hosts diverse microbial community. Ground Water. 2006;44:511–7.
[100] Krulwich TA, Sachs G, Padan E. Molecular aspects of bacterial pH sensing and homeostasis Nat Rev Microbiol. 2011;9:330–43.
[101] Horikoshi K. Alkaliphilies: some applications of their products for biotechnology. Microbiol Molecul Biol Rev. 1999;63:735–50.
[102] Krulwich TA, Hicks DB, Swartz TH, Ito M. Bioenergetic adaptations that support alkaliphily. In: Gerday C, Glansdorff N, editors. Physiology and biochemistry of extremophiles. Washington, DC: ASM Press; 2007. p. 311–329.
[103] Slonczewski JL, Fujisawa M, Dopson M, Krulwich TA. Cytoplasmic pH measurement and homeostasis in bacteria and archaea. Adv Microb Physiol. 2009;55(1–79):317.

[104] Zhao Y, Zhang Y, Cao Y, et al. Structural analysis of alkaline β-mannanase from alkaliphilic *Bacillus* sp. N16-5: implications for adaptation to alkaline conditions. PLoS ONE. 2011;6(1):e14608.
[105] Fujinami S, Fujisawa M. Industrial applications of alkaliphiles and their enzymes – past, present and future. Environ Technol. 2010;31:845–56.
[106] Mamo G, Mattiason B. Alkaliphilic microorganisms in biotechnology. In: Rampelotto PH, editor. Biotechnology of extremophiles: advances and challenges (grand challenges in biology and biotechnology). New York, NY: Springer; 2016. 243–72.
[107] Andreaus J, Olekszyszen DN, Silveira MHL. Processing of cellulosic textile materials with cellulases in cellulose and other naturally occurring polymers. In: Fontana JD, Tiboni M, Grzybowski A, editors. Cellulose and other naturally occurring polymers. Kerala, India: Research Signpost; 2014. p. 11–19.
[108] Van den Burg B. Extremophiles as a source for novel enzymes. Curr Opin Microbiol. 2003;6:213–8.
[109] Fang J, Zhang L, Bazylinski DA. Deep-sea piezosphere and piezophiles: geomicrobiology and biogeochemistry. Trends Microbiol. 2010;18(9):413–22.
[110] Zhang Y, Li X, Bartlett DH, Xiao X. Current developments in marine microbiology: High-pressure biotechnology and the genetic engineering of piezophiles. Curr Opin Biotechnol. 2015;33:157–64.
[111] Michoud G. Jebbar M. High hydrostatic pressure adaptive strategies in an obligate piezophile *Pyrococcus yayanosii*. Sci Rep. 2016;6:27289.
[112] Kato C. Cultivation methods for piezophiles. In: Horikoshi K, Antranikian G, Bull A, Robb F, Stetter K, editors. Tokyo, Japan: Springer-Verlag; 2011. p. 719–26.
[113] Birrien JL, Zeng X, Jebbar M, et al. *Pyrococcus yayanosii* sp. nov., an obligate piezophilic hyperthermophilic archaeon isolated from a deep-sea hydrothermal vent. Int J Syst Evol Microbiol. 2011;61:2827–31.
[114] Tamegai H, Nishikawa S, Haga M, Bartlett DH. The respiratory system of the piezophile *Photobacterium profundum* SS9 grown under various pressures. Biosci Biotechnol Biochem. 2012;76:1506–10.
[115] Amrani A, Bergon A, Holota H, et al. Transcriptomics reveal several gene expression patterns in the piezophile *Desulfovibrio hydrothermalis* in response to hydrostatic pressure. PLoS ONE. 2014;9:e106831.
[116] Abe F, Horikoshi K. The biotechnological potential of piezophiles. Trends Biotechnol. 2001;19:3.
[117] Zhang Y, Li X, Bartlett DH, Xiao X. Current developments in marine microbiology: high-pressure biotechnology and the genetic engineering of piezophiles. Curr Opin Biotechnol. 2015;33:157–64.
[118] Singh OV, Gabani P. Radiation resistance microbial reserves and therapeutic implications. J Appl Microbiol. 2011;110:851–61.
[119] Pavlopoulou A, Savva GD, Louka M, Bagos PG, Vorgias CE, Michalopoulos I, Georgakilas AG. Unraveling the mechanisms of extreme radioresistance in prokaryotes: Lessons from nature. Mutat Res Rev. 2016;767:92–107.
[120] Pikuta EV, Hoover RB, Tang J. Microbial extremophiles at the limits of life. Crit Rev Microbiol. 2007;33:183–209.
[121] Webb KM, DiRuggiero J. Role of Mn^{2+} and compatible solutes in the radiation resistance of thermophilic Bacteria and Archaea. Archaea. 2012;845756.
[122] Daly MJ. A new perspective on radiation resistance based on *Deinococcus radiodurans*. Nat Rev Microbiol. 2009;7:237–45.
[123] Daly MJ, Gaidamakova EK, Matrosova VY, Kiang JG, Fukumoto R, Lee D-Y, Wehr NB, Viteri GA, Berlett BS, Levine RL. Small-molecule antioxidant proteome-shields in *Deinococcus radiodurans*. PLoS ONE. 2010;5:e12570.

[124] Albarracín V, Gärtner W, Farias ME. Forged under the sun: life and art of extremophiles from Andean lakes. Photochem Photobiol. 2016;1:14–28.

[125] Nicholson WL, Fajardo-Cavazos P, Rebeil R, et al. Bacterial endospores and their significance in stress resistance. Antonie van Leeuwenhoek. 2002;81:27–32.

[126] Farci D, Bowler MW, Kirkpatrick J, McSweeney S, Tramontano E, Piano D. New features of the cell wall of the radio-resistant bacterium *Deinococcus radiodurans*. Biochim Biophys Acta. 2014;1838(7):1978–84.

[127] Dalmaso GZL, Ferreira D, Vermelho AB. Marine extremophiles: a source of hydrolases for biotechnological applications. Mar Drugs. 2015;13:1925–65.

[128] Dziewit L, Grzesiak J, Ciok A, Nieckarz M, Zdanowski MK, Bartosik D. Sequence determination and analysis of three plasmids of *Pseudomonas* sp. GLE121, a psychrophile isolated from surface ice of Ecology Glacier (Antarctica). Plasmid. 2013;70(2):254–62.

[129] Turick CE, Ekechukwu AA, Milliken CE, Casadevall A, Dadachova E. Gamma radiation interacts with melanin to alter its oxidation-reduction potential and results in electric current production. Bioelectrochemistry. 2011;82:69–73.

[130] Gabani P, Singh OV. Radiation-resistant extremophiles and their potential in biotechnology and therapeutics. Appl Microbiol Biotechnol. 2013;97:993–1004.

[131] Kim SJ, Koh DC, Park SJ, et al. Molecular analysis of spatial variation of iron-reducing bacteria in riverine alluvial aquifers of the Mankyeong River. J Microbiol. 2012;50(2):207–17.

[132] Seyrig G. Uranium bioremediation: current knowledge and trends. Basic Biotechnol eJournal. 2010;6(1):3.

[133] Liao CS, Chen LC, Chen BS, Lin SH. Bioremediation of endocrine disruptor di-n-butyl phthalate ester by *Deinococcus radiodurans* and *Pseudomonas stutzeri*. Chemosphere. 2010;78(3):342–6.

[134] Dalmaso GZL, Lage CAS, Mazotto AM, et al. Extracellular peptidases from *Deinococcus radiodurans*. Extremophiles. 2015;19:989–99.

[135] Tian B, Sun Z, Shen S, et al. Effects of carotenoids from *Deinococcus radiodurans* on protein oxidation. Lett Appl Microbiol. 2009;49(6):689–94.

[136] Coker JA. Extremophiles and biotechnology: current uses and prospects. F1000Research. 2016;5.

[137] Krüger A, Scäfers C, Schöder C, Antranikian G. Towards a sustainable biobased industry – highlighting the impact of extremophiles. New Biotechnol. 2018;40(25):144–53.

[138] Kumar A, Singh S. Directed evolution: tailoring biocatalysts for industrial applications. Crit Rev Biotechnol. 2013;33(4)365–78.

[139] Reetz MT. Laboratory evolution of stereoselective enzymes: a prolific source of catalysts for asymmetric reactions. Angew Chem Int Ed. 2011;50:138.

[140] Damian-Almazo JY, Saab-Rincon G. Site-directed mutagenesis as applied to biocatalysts. In: Figurski D, editor. Biochemistry, genetics and molecular biology. Rijeka: InTech; 2013; 450p.

[141] Shu ZY, Wu JG, Xue LY, et al. Construction of *Aspergillus niger* lipase mutants with oil-water interface independence. Enzyme Microb Technol. 2011;48:129–33.

[142] Xu H, Qin Y, Huang Z, Liu Z. Characterization and site-directed mutagenesis of an α-galactosidase from the deep-sea bacterium *Bacillus megaterium*. Enzyme Microb Technol. 2014;5(56):46–52.

[143] Yasukawa K, Inouye K. Improving the activity and stability of thermolysin by site-directed mutagenesis. Biochim Biophys Acta. 2007;1774(10):1281–8.

[144] Chi MC, Chen YH, Wu TJ, Lo HF, Lin LL. Engineering of a truncated α-amylase of *Bacillus* sp. strain TS-23 for the simultaneous improvement of thermal and oxidative stabilities. J Biosci Bioeng. 2010;109:531–8.

[145] Lidan Y, Hua Z, Zhi L, Wu JC. Improved acid tolerance of *Lactobacillus pentosus* by error-prone whole genome amplification. Bioresour Technol. 2013;135:459–63.
[146] Zhang N, Suen WC, Windsor W, Xiao L, Madison V, Zaks A. Improving tolerance of *Candida antarctica* lipase B towards irreversible thermal inactivation through directed evolution. Protein Eng. 2003;16:599–605.
[147] Liu Y, Zhang T, Zhang Z, Sun T, Wang J, Lu F. Improvement of cold adaptation of *Bacillus alcalophilus* alkaline protease by directed evolution. J Mol Catal B Enzym. 2014;106:117–25.
[148] Park H, Ahn J, Lee J, et al. Expression, immobilization and enzymatic properties of glutamate decarboxylase fused to a cellulose-binding domain. Int J Mol Sci. 2012;13(1):358–68.
[149] Chakraborty S, Khopade A, Kokare C, Mahadik K, Chopade B. Isolation and characterization of novel α-amylase from marine *Streptomyces* sp. D1. J Mol Catalysis B Enzymatic. 2009;58:17–23.
[150] Hayashi R. Use of high pressure in bioscience and biotechnology. In: Hayashi R, Balny C, editors. High pressure bioscience and biotechnology. London: ENG, Elsevier; 1996; 521p.
[151] Nisha M, Satyanarayana T. Recombinant bacterial amylopullulanases: developments and perspectives. Bioengineered. 2013;4(6):388–400.
[152] Alias N, Ahmad Mazian M, Salleh AB, Basri M, Rahman RNZRA. Molecular cloning and optimization for high level expression of cold-adapted serine protease from Antarctic yeast *Glaciozyma antarctica* PI12. Enzyme Res. 2014;197938.
[153] Ferrer M, Chernikova TN, Yakimov MM, Golyshin PN, Timmis KN. Chaperonins govern growth of *Escherichia coli* at low temperatures. Nat Biotechnol. 2003;21(11):1266–7.
[154] Khan I, Kihara D. Computational characterization of moonlighting proteins. Biochem Soc Trans. 2014;42:1780–5.
[155] Galperin MY, Koonin EV. From complete genome sequence to 'complete' understanding? Trends Biotechnol. 2010;28(8):398–406.
[156] Jiao WB, Accinelli GG, Hartwig B, et al. Improving and correcting the contiguity of long-read genome assemblies of three plant species using optical mapping and chromosome conformation capture data. Genome Res. 2017;27(5):778–86.
[157] Simon C, Daniel R. Metagenomic analyses: past and future trends. Appl Environ Microbiol. 2011;77(4):1153–61.
[158] Ferrer M, Martínez-Martínez M, Bargiela R, Streit WR, Golyshina OV, Golyshin PN. Estimating the success of enzyme bioprospecting through metagenomics: current status and future trends. Microb Biotechnol. 2016;9:22–34.
[159] Gupta SK, Shukla P. Advanced technologies for improved expression of recombinant proteins in bacteria: perspectives and applications. Crit Rev Biotechnol. 2015;18:1–10.
[160] Mckenzie GJ, Craig NL. Fast, easy and efficient: site-specific insertion of transgenes into Enterobacterial chromosomes using Tn*7* without need for selection of the insertion event. BMC Microbiol. 2006;6:39.
[161] Wong D. Metagenomics: current advances and emerging concepts. In: Marco D, editor. Metagenomics: theory, methods and applications. Cordoba, Argentina: Caister Academic Press; 2010; 226p.
[162] Suenaga H. Targeted metagenomics unveils the molecular basis for adaptive evolution of enzymes to their environment. Front Microbiol. 2015;6:1018.
[163] Alma'abadi AD, Gojobori T, Mineta K. Marine metagenome as a resource for novel enzymes. Genomics Proteomics Bioinformatics. 2015;13(5):290–5.
[164] Kennedy J, Flemer B, Jackson SA, et al. Marine metagenomics: new tools for the study and exploitation of marine microbial metabolism. Mar Drugs. 2010;8(3):608–28.
[165] Ferrandi EE, Sayer C, Isupov MN, et al. Discovery and characterization of thermophilic limonene-1,2-epoxide hydrolases from hot spring metagenomics libraries. FEBS J. 2015;282:2879–94.

[166] Lopez-Lopez O, Cerdan ME, Gonzalez-Siso MI. New extremophilic lipases and esterases from metagenomics. Curr Protein Pept Sci. 2014;15:445–55.
[167] Sharma VK, Kumar N, Prakash T, Taylor TD. MetaBioME: a database to explore commercially useful enzymes in metagenomic datasets. Nucleic Acids Res. 2010;38(1):468–72.
[168] Gaida SM, Sandoval NR, Nicolaou SA, Chen Y, Venkataramanan KP, Papoutsakis ET. Expression of heterologous sigma factors enables functional screening of metagenomic and heterologous genomic libraries. Nat Commun. 2015;6:7045.
[169] Burg D, Ng C, Ting L, Cavicchioli R. Proteomics of extremophiles. Environ Microbiol. 2011;13:1934–55.
[170] Yun SH, Choi CW, Lee SY, Park EC, Kim SI. A proteomics approach for the identification of novel proteins in extremophiles. In: Rampelotto PH, editor. Biotechnology of extremophiles: advances and challenges (grand challenges in biology and biotechnology). New York, NY: Springer; 2016. p. 303–19.
[171] Yun SH, Choi CW, Kwon SO, et al. Enrichment and proteome analysis of a hyperthermostable protein set of archaeon *Thermococcus onnurineus* NA1. Extremophiles. 2011;15:451–61.
[172] Antranikian G, Vorgias CE, Bertoldo C. Extreme environments as a resource for microorganisms and novel biocatalysts. Adv Biochem Eng Biotechnol. 2005;96:219–62.
[173] Chen YY, Galloway KE, Smolke CD. Synthetic biology: advancing biological frontiers by building synthetic systems. Genome Biol. 2012;13:240.
[174] Otero JM, Cimini D, Patil KR, Poulsen SG, Olsson L, Nielsen J. Industrial systems biology of *Saccharomyces cerevisiae* enables novel succinic acid cell factory. PLoS ONE. 2013;8:e54144.

Mia M. Howard, Laura M. Kaminsky, André Kessler
and Terrence H. Bell

4 Merging microbial and plant profiling to understand the impact of human-generated extreme environments on natural and agricultural systems

Abstract: Many human activities, such as agriculture and pollution, have severely disturbed soil habitats, exposing resident plants and microorganisms to relatively extreme environmental conditions. This chapter discusses the mechanisms through which anthropogenic disturbances can create extreme environments for soil organisms, with a particular focus on the disruption of plant-microbe interactions. We then provide recommendations for studying the impact of such disturbances on plant-microbe interactions through combined microbial and plant profiling. We highlight decision-making processes for designing both field surveys and manipulative experimental studies, such as microbiome transplant experiments, to investigate the impacts of such disturbances. This is done through the lens of a theoretical study examining the effects of agricultural cultivation on plants and their associated microbial communities.

4.1 Generation of extreme environments through human activity

Many ecosystems have been altered dramatically as a result of human activity [1, 2]. Both intentional (e.g. agriculture) and unintentional (e.g. pollution) human alterations of plant systems can create "extreme" environments for plants and microorganisms, particularly in the soil. The biotic and abiotic constraints of a habitat can be rapidly and severely shifted in a way that makes it difficult for the native (i.e. pre-disturbance) community to persist. Depending on the nature and intensity of the disturbance, these environmental shifts could lead to community dominance by organisms with entirely different life strategies than those found pre-disturbance [3].

In this chapter, our definition of *extreme* is relative to pre-disturbance plant systems and does not require conditions that push the limits of what is tolerable to most forms of life on Earth [4]. While certain human disturbances can indeed create conditions that are inhospitable to most organisms, such as the extremely alkaline conditions created by radioactive waste disposal [5] or the extreme heat, acidity, and heavy metal concentrations of waste from coal mines [6], most human-influenced plant systems are not overly hostile to life. Instead, the most common human disturbances alter the environment in ways that reduce the competitiveness of the pre-disturbance

https://doi.org/10.1515/9783110525786-004

community, even when the physiology of individual species could allow them to survive in the absence of competition. These disruptions can substantially alter the structure and composition of communities by the relative exclusion and promotion of different taxa and by altering species interactions [7], potentially leading to large shifts in system function over time. Furthermore, many human disturbances, unlike natural disturbances, often occur with unpredictable temporal variability, rather than reflecting historical cycles, and expose organisms to conditions outside the realm of those in which they have evolved [1]. Such disturbances can therefore create novel environments, which may favor entirely different, opportunistic species and genotypes [1].

Many human disturbances to natural habitats, such as agriculture and chemical pollutants, manifest in soils, and thus, the effects of disturbances on the biotic assemblages of soils are the focus of this chapter. Changes in soil microbial communities are likely to affect the plants with which they associate, as plant communities and phenotypes can be affected by their interactions with microorganisms [8, 9]. Understanding how disturbances affect soil microbes, and consequently plants, can be particularly important in human-altered plant systems, which we actively cultivate, such as agricultural fields and silvicultures. Dramatic alterations of microbe-plant interactions, as can occur in those systems, are known to affect important ecosystem functions such as mobilization of nutrients and nutrient cycling, which may impact the sustainability of cultivation practices [10] and global nutrient cycles. Nevertheless, plant breeding efforts have not traditionally targeted the optimization of plant-microbe relationships per se, mainly because of the limited understanding of the complex interrelations of microbes and plants.

This chapter discusses the study of human disturbances on soil habitats, with a focus on the disruption of plant-microbe relationships. The first section examines the major mechanisms through which human activities create extreme environments in the soil habitat, and provides examples of commonly disturbed plant systems. In the second section, we provide recommendations for examining these disturbed systems through combined microbial and plant profiling.

4.1.1 Examples of human disturbance to plant systems

This section provides examples of common plant systems that have been altered by human activity and describes the primary modes through which disturbances are expected to make the soil habitat an extreme environment for resident plants and microorganisms. Many human-altered ecosystems have been disturbed in multiple ways (Tab. 4.1). Because they are essentially sessile, plants and their associated microorganisms are likely to be particularly affected by environmental disturbances. It is also important to note that both plants and microorganisms, as well as different species of each, are likely to be uniquely affected by different types of disturbances, due to differences in their physiology and levels of tolerance for various stresses. Moreover, these "disturbances" may not all have negative effects on plant and microbial communities or plant-microbe relationships. For example, fertilizer application could promote both plant and microbial growth, as well as reduce negative plant-microbe interactions by relaxing competition between plants and microbes for nutrients [11].

Tab. 4.1: Examples of human-disturbed ecosystems and their primary modes of disturbance.

	TYPE OF DISTURBANCE				
HUMAN-ALTERED PLANT SYSTEM	**Physical disruption**	**Nutrient additions**	**Nutrient removal**	**Chemical changes/ additions**	**Inhibitory organisms**
Natural plant systems	–	Deposition of air pollutants[18]	–	Deposition of air pollutants[18]; changes in soil pH[33–35]; runoff from nearby managed systems	Migration of organisms from nearby managed systems
Managed forests	Erosion from logging roads/ activities; compaction from vehicle traffic[72]	Deposition of air pollutants[18]	Tree and brush removal[63]; soil erosion	Deposition of air pollutants[18]; changes in soil pH[33–35]; Vehicle exhaust[36]	–
Roadsides	–	Deposition of air pollutants[18]	Mowing[63]	Vehicle exhaust[16]; road chemical runoff; changes in soil pH[33–35]	–
Recreational fields	–	Deposition of air pollutants[18]	Mowing[63]	Changes in soil pH[33–35]; Herbicides; pesticides; fungicides[48–52, 54]	–
Lawns	–	Fertilizer amendments[11, 19–25]	Mowing[63]	Changes in soil pH[33–35]; Herbicides; pesticides; fungicides[48–52, 54]	–
Farms, Orchards, and Gardens	Compaction from vehicle, animal, and foot traffic [72–74]; tillage[64, 65, 77, 79, 80]; weeding; irrigation; soil erosion	Fertilizer amendments[11, 19–25]	Harvesting[26]; weeding; soil erosion	Changes in soil pH[33–35]; Herbicides; pesticides; fungicides[48–52, 54]	Increase in crop-specific pathogens [86–88]; pesticide-tolerant organisms

Altering soil nutrients

Nutrient cycling is one of the most critical ecosystem services that microorganisms provide in the soil. Soil nutrients – particularly nitrogen, phosphorus, and potassium – fulfill essential biological roles for both plants and microbes, and as such, the availability of these nutrients plays a major role in determining which organisms can survive and thrive in a given environment. In undisturbed systems, the availability of these nutrients for plant or microbial uptake is driven in large part by microbes, who are key players in these biogeochemical cycles. For example, microbes are responsible for many of the chemical transformations in the soil nitrogen (N) cycle, converting organic N, and in special cases atmospheric N_2, to and from the inorganic, plant-available forms nitrate (NO_3^-) and ammonium (NH_4^+) [12]. Likewise, microbial activity drives the soil phosphorous cycle, solubilizing fixed inorganic P through mechanisms such as the release of organic acids [13] and mineralizing fixed organic P through the release of extracellular enzymes such as phosphatases and phytases into plant-available phosphate ions (PO_4^{3-}) [14]. Additionally, certain microbes (e.g. arbuscular mycorrhizal fungi) can enhance the ability of plants to acquire nutrients by forming symbiotic relationships with roots and scavenging P and other nutrients for the plant through hyphal networks that extend far beyond the reach of root systems [15, 16]. Thus, disruptions to soil nutrient cycling, whether through altering amounts of nutrients, or directly disturbing microbes, is likely to cause substantial alterations to the composition and function of soil microbial communities.

It is important to remember that while microbes convert nutrients to plant-available forms, they also require these nutrients for their own growth and metabolism and can thus directly compete with plants for these resources [17]. Much of the N or P that is released by microbial activity is subsequently taken up by the microbes themselves and is consequently unavailable for plants until the microbial cells lyse [13, 15]. Human disturbances that substantially alter nutrient pools and flow may contribute to the creation of extreme environments by promoting competitive conditions that are atypical for the site, disfavoring some established microorganisms.

Adding nutrients: To boost yields and maintain productivity, nutrients are regularly added to managed plant-soil systems (e.g. agricultural fields and turfgrass) through the use of organic (e.g. compost and manure) or chemical fertilizers. These usually constitute formulations of N, P, and K in various forms and ratios. Nutrient inputs may also be indirect and unintentional. For example, N deposition to terrestrial ecosystems has increased since the Industrial Revolution, as the burning of fossil fuels releases more nitrogen oxides into the atmosphere that are then washed into soil during rain [18]. These inputs, by disproportionately and artificially enhancing certain soil nutrient pools, change the overall stoichiometry of the soil, differentially favoring groups of microbes that are adapted to different nutrient conditions [19]. For example, certain nutrient acquisition traits may no longer be advantageous when those nutrients are readily available, and so microbes that exhibit these previously useful traits are likely to be outcompeted by weedier species that do not invest as much in accessing recalcitrant nutrient pools [20].

Many studies in plant systems, ranging from grasslands to paddy soils, have robustly demonstrated that microbial communities respond to nutrient additions [19–23]. Even across these different systems and with widely varying fertilization regimes, all have found significant changes in microbial community structure following nutrient additions, with some overlapping patterns. For instance, different fertilizer types (inorganic mineral fertilizers vs. composts, manures, and other organic additions) have markedly different effects on the soil microbiome [20, 22, 23], although the exact soil parameter thought to be driving these changes ranges from organic C content in one study [22] to ammonium content in another [23]. Most studies have found that alterations in the relative abundance of bacterial and fungal taxa can be observed at high taxonomic levels (e.g. phylum) under different nutrient regimes [19–23], although in some cases, responses are more subtle (e.g. changes in operational taxonomic units) and are not as evident at higher taxonomic levels [21, 22]. Despite these discrepancies, clearer trends emerge when the same nutrient additions are applied to soils worldwide; for example, under high nutrient conditions, there is an increase in putatively copiotrophic microbes relative to oligotrophic microbes [29].

Certain keystone nutrient cycling microbes have been studied in greater detail and can be strongly affected by soil nutrient changes, particularly when fertilization reduces the dependency of plants on microbial symbionts for nutrient acquisition. For example, rhizobial abundance and the rate of biological nitrogen fixation in legume roots decline with greater N availability [24]. The abundance of arbuscular mycorrhizal fungi can decline as soil P fertility increases, since it is no longer as advantageous for plants to engage in the fungal symbiosis under high P conditions [16]. Phosphorus-solubilizing microbes can also decline in abundance under conditions of higher P availability [25].

Depleting nutrients: Conversely, if nutrient pools are not replenished in systems managed for crop production, soil nutrient depletion can occur, forcing the community to cope with a shift from nutrient abundance (under-fertilization) to scarcity. For example, a typical corn harvest in the United States removes 132.8 kg N ha^{-1}, 39 kg P ha^{-1}, and 49 kg K ha^{-1} from the soil [26], and other staple crops are similarly demanding [27]. If these harvested nutrients are not replaced with organic or inorganic amendments, soil nutrient pools can quickly become exhausted. This problem plagues over 60% of agricultural land worldwide [27]. In such soils, microbes with the capability to acquire nutrients from more recalcitrant nutrient pools, and which were not likely to have been at a competitive advantage when fertilizer was applied, could increase once again in abundance [28]. Plants may also invest more resources into recruiting microbes that help them acquire limiting nutrient resources, or release root exudates that suppress microbes that compete with them for limited nutrients [29]. Again, these microbial adjustments under nutrient stress can significantly alter the overall soil microbiome, which in turn can impact plant performance.

pH modification

Soil pH is another soil property frequently altered by human activity. This alteration can be intentional; for example, most agricultural crops thrive at a mid-range soil pH of 6–7; thus, soils with a natural pH outside this range are often amended with lime or sulfur to raise or lower the pH as required. Soil pH may also be changed indirectly (e.g. through fertilizer addition). Ammonium fertilizers in particular tend to lower pH, since plants must release protons to acquire these cations from the soil [30]. Unintentional changes in soil pH can also occur via acid rain or acid mine drainage [31, 32].

Changes in soil pH are a strong driving force behind microbial community composition, more so than any other ecosystem property [33]. This may be because alterations in pH can impact many other soil properties in tandem, such as the concentration and availability of nitrate, ammonium, phosphate, and micronutrients [21], as well as soil moisture and organic carbon content [34]. Nonetheless, soils with different pH values tend to have different microbial communities, even when controlling for other soil factors. For example, bacterial abundance and community composition were strongly linked with pH across a continuous 180-m-long field that ranged from a pH of 4 to a pH of 8, in which most other soil parameters were comparable [34]. The microbial differences from one end of this pH-gradient to the other were as dramatic as in a global study linking soil pH and microbiome structure [34, 35]. Human influence on pH can therefore create an extreme environment that shapes the microbial community, which may then have a secondary effect on plant performance in addition to the plant's direct response to soil pH.

Chemical contamination of soils

Pollution: Many habitats are unintentionally altered by humans through exposure to pollutants. Soil contamination can arise from a point-source (e.g. oil spills) or from a variety of diffuse sources (e.g. car exhaust) [36]. Many common contaminants have direct negative effects on both microorganisms and plants. For example, arsenic, zinc, copper, cadmium, and vanadium have all been observed to decrease soil microbial biomass and functional diversity in exposed soil communities [37–39], and heavy metals can also interfere with plant photosynthesis and water retention [40, 41]. Hydrocarbons have been shown to negatively impact bacterial membrane permeability, proteins, and lipid composition [42]; damage plant cell membranes; and reduce transpiration and translocation [43].

Some microorganisms can assist with plant tolerance of soil contaminants [44], as well as other human-impacted abiotic stressors, such as salt [45]. This service implies superior stressor tolerance by the involved microorganisms, relative to the plant, which could potentially shift the balance of interdependence between plants and their associated microorganisms (i.e. "I need you more than you need me"). Little is known about shifts in the power dynamics of plant-microbe relationships under stress, although plants have been shown to allocate more C to root exudates under stressful conditions [46]. Contaminants will also vary in their impact on different microbial groups [47]. This alters the available pool of plant-associated microorganisms and, even without direct effects on plants, alters the nature of plant-microbe relationships.

Agricultural chemicals: Modern agricultural practices rely heavily on the input of synthetic compounds designed to treat specific pest problems. The use of agricultural chemicals such as pesticides, fungicides, and herbicides can unintentionally alter communities of microorganisms in the soil, with potentially negative effects on soil and plant health. Firstly, many pesticides, such as organophosphates, can be used as microbial energy sources [48], and the application of such chemicals could shift microbial communities to favor microbes that can metabolize this new carbon source, especially when other carbon sources are limited. Secondly, agricultural chemicals can have toxic effects on microbes. Since the mode of action of many pesticides and herbicides involves the inhibition of enzymes in their target organism(s), these chemicals can also disrupt microbial enzymes in the soil [49]. Consequently, even agricultural chemicals not specifically designed to target microbes, such as insecticides and herbicides, can disrupt communities of soil bacteria and fungi [50, 51].

At the most extreme end are chemicals used to sterilize soils. Soils are often chemically fumigated for the purpose of eliminating pathogenic microorganisms and nematodes, but fumigants broadly kill soil organisms, reducing the abundance of mutualistic microbes such as mycorrhizal fungi, in addition to pathogens [52]. While soil sterilization can improve plant growth through killing antagonistic organisms and releasing nutrients such as mineralized N [53, 54], it causes shifts in indigenous microbial communities and has been observed to reduce both functional and taxonomic diversity in treated soils [55, 56]. Alternatives to chemical fumigation for sterilizing soil, such as solarization, also create extreme environments for microbes (e.g. extreme heat in the case of solarization) and have been found to alter the composition of microbial communities [57]. The recolonization process of sterilized soil in the field is not completely understood, but recent *in vitro* studies, in which sterilized soils were reinoculated with their previous microbial communities, found that sterilized soils were colonized by many taxa that were not detected prior to sterilization, suggesting promotion of rare microorganisms [58] and that the new microbial communities are likely to be functionally different [59]. The removal of competitors during sterilization could potentially create opportunities for rare or introduced microbes, such as invading pathogens or mycorrhizal inoculants [60], to establish sizable populations, as well as for opportunistic species that can take advantage of the unoccupied habitat and flush of nutrients released from organisms killed by fumigation [53, 54].

Physical soil disturbances

In addition to chemical disturbances, physical disruptions to the soil from human activities can reshape the soil habitat and create extreme environments for microorganisms. Many human-altered habitats are disrupted by physical disturbances, such as tillage and the uprooting of plants through weeding or harvesting, which modify the soil environment in numerous ways.

Organic matter depletion: Soil organic matter, the primary source of energy for soil microorganisms and a major reservoir of nutrients and moisture, is often depleted in human-altered plant systems [61]. The amount of soil organic carbon is a strong predictor of microbial biomass [62], and consequently, depletion of this resource could severely reduce microbial abundance. Organic matter is depleted when biomaterials such as plants are removed (e.g. harvested) but not returned to the soil. Thus, the harvesting of crops and removal of weeds and crop residues reduce organic substrates that, in an unmanaged system, would be decomposed and would help sustain portions of the soil community [63]. Soil organic matter can also be degraded during tillage, or other processes that mix or disrupt the soil profile, which promotes erosion and speeds the rate of organic matter decomposition by increasing soil temperature and oxygenation [64, 65].

Altering soil texture: Soil aggregates, or groups of bound soil particles, create a structure of pore spaces in the soil that help regulate soil moisture and aerobicity [66]. Because the availability of both water and oxygen influences soil microbial community composition and activity [67–69], the physical structure of the soil can greatly impact soil biology. The stability of soil aggregates is dependent on microbes, especially fungi, that can bind soil aggregates physically, as well as chemically using compounds such as sugars generated during the decomposition of organic matter [66, 70]. Thus, there is a feedback in which changes to microbial communities can affect soil texture, which in turn impact microbial communities through altered soil relationships with air and water. Fungal communities, in particular, have been found to be closely related to soil aggregate stability [71]. The texture of soils can be disrupted by physical breakage and compaction of soil aggregates, through disturbances such as tillage and the use of other heavy machinery [65, 72], as well as heavy traffic on top of the soil by humans [73] and farm animals [74].

Disrupting fungal hyphal networks: Physical disturbances to the soil, including tillage, can directly affect soil fungi by physically breaking the hyphal connections that fungi use for nutrient and water absorption, as well as propagation [75, 76]. Tillage in maize fields has been found to reduce the density of mycorrhizal hyphae in the soil by 20–27%, depending on soil type, resulting in reduced mycorrhizal colonization of maize roots [77]. In a meta-analysis, intensive soil tillage was found to be associated not only with decreased mycorrhizal colonization of crop roots but also with decreased diversity of colonizing fungi [78]. Fungal hyphae may also play an important role in stabilizing soil aggregates [66]. Thus, physical disturbances, such as tillage or root harvesting, can make the soil habitat less hospitable for microorganisms, not only directly by disrupting fungi through damage to their hyphal networks but also indirectly by altering soil properties previously managed by hyphae.

Altering soil layers: Physical disruptions of the soil profile during processes such as soil turning during tillage can force microbes into inhospitable environments. The physical characteristics of soil layers vary greatly, and even on a microscopic scale, there can be substantial differences in environmental conditions such as pH, moisture, oxygen level, and nutrients [79, 80]. Thus, even moving microbes a small distance in the soil could put them in a drastically different environment. During tillage, for example, anaerobic microbes buried beneath the surface of the soil could suddenly be moved to the oxygenated surface, and vice versa. Tilled soils have been found to have proportionately fewer anaerobes than their no-till counterparts in the upper soil layers, as well as lower potential for denitrification (a type of anaerobic respiration), indicating that such disturbances can affect microbial composition, with functional effects on ecosystem services such as N cycling [68].

Promoting accumulation of pathogens

Human activities, particularly those that reduce plant and microbial diversity, can promote plant diseases. It has been widely observed that the rates and severity of plant diseases are high when crops are planted in monoculture [81, 82], particularly genetic monoculture [83–85], as compared to more diversely planted agricultural systems. Repeatedly planting the same crops in the same soil can cause "crop replant disease" in some cases, as the availability of the same hosts in high densities can promote the accumulation of pathogens that are specific to those hosts [86–88]. In natural systems, the relationship between plant diversity and plant productivity [89, 90] and disease resistance [89, 91] is at least partially mediated by soil microorganisms, likely due to decreased loads of pathogenic bacteria and fungi in the soil supporting diverse plant communities.

4.1.2 Microbes affect plant communities and phenotypes

Examining microbial community shifts is an important part of understanding how human disturbances impact plants, as microbes and plants are intrinsically linked on multiple levels of organization, from cells to whole communities. Soil microbial diversity has been observed to be correlated with the diversity of the aboveground plant community in both natural and human altered systems [92, 93], and increased microbial diversity may at least partially explain why diverse plant communities tend to be highly productive [89]. There is evidence that a soil community can, to some extent, beget the plant community, determining which plants colonize and thrive. For example, during the restoration of ecosystems, inoculating soil with heathland versus grassland soil communities promoted the establishment of plants from the respective habitats [8]. Thus, due to the close relationships that microbes form with plants, changes to soil microbial communities can have major effects on the composition of plant communities.

Likely underlying the observed effects on plant community composition, microorganisms can affect plant growth, performance, and metabolism. In fact, the plant microbiome has been called the second genome of plants because interactions with associated microbes have been found to affect many plant traits [9]. Associations with mycorrhizal fungi [94] and symbiotic bacteria [95], for example, have been shown to enhance plant growth, and consequently, humans often attempt to modify plant microbiomes in agriculture through the use of bioinoculants, as a means of improving plant performance [96–98]. Soil microbial communities can also affect plant phenology and fitness [99, 100] and the biochemistry of endophytic microbes can supplement the metabolism of plants, expanding the range of chemicals they can produce [101]. Such effects on plant secondary metabolism can, in turn, affect plant interactions with herbivores [102–104], pathogens [105], and other plants [106]. The targeted manipulation of plant microbiomes improve plant performance is a promising area of research [107] and novel methods of microbiome modification, including that of the seed, have recently been developed [108].

The strong interdependence of plants and microbes seems to be mirrored in interactions of microbes with other macroorganisms, painting a picture of apparent ubiquity in mutualist symbiosis. However, there are examples of macroorganisms that do not appear to depend on symbiotic interactions with microbes. For example, preliminary studies suggest that some insects such as walking sticks [109] and lycaenid caterpillars [110] might not need bacterial symbionts in their guts to digest plant tissue, a trait initially assumed to be ubiquitous to all herbivores. Similarly, while most plant species support mycorrhizal symbioses, some plants, such as those in the Brassicaceae family, do not typically form these relationships [111]. Thus, as the extent of plant-microbe relationships is still being characterized, it should not be assumed that all plants are equally dependent on microbial interactions. It may be that relative dependence on microbial symbionts could differ based on the circumstances under which the plants live. For example, plants that regularly tolerate higher stress conditions, such as extreme heat [112], may tend to be more microbially dependent than plants in environments without these stressors. Similarly, there may be a relationship between life history strategies and microbial dependence, with weedier, opportunistic species potentially requiring less microbial support than later successional species that compete more intensely for nutrients [113].

4.1.3 Plant breeding and plant-microbe relationships

Traditional plant breeding under modern agricultural conditions has focused on plants as independent organisms, rather than as holobionts (plants and their microbiomes as an interdependent whole), which may not only result in less resilient crops [114] but also in further disruption of plant-associated microbial communities in human-managed systems. Because crop plants are typically grown under conditions that aim to reduce the stress on individuals and minimize biotic interactions and dependence on soil nutrient cycles (e.g. high nutrient availability through fertilizer, supplemental irrigation, and release from herbivory and pathogens through pesticides), it has been suggested that cultivated plants may engage in fewer stress-mitigating interactions with microorganisms than their wild relatives do, as these may not have been selected for through traditional plant breeding strategies [115]. For example, the benefits to plants of establishing relationships with mycorrhizal fungi can decline (and can even become detrimental/parasitic to the plant) under high nutrient conditions [116, 117], suggesting that the ability of plants to form beneficial relationships with arbuscular mycorrhizal fungi could decrease when bred in soil supplemented with fertilizer; however, this overall trend has not been definitively observed across crops [118]. Similarly, modern soybean varieties bred under soil conditions with adequate nitrogen have been found to be less able to selectively establish effective relationships with nitrogen-fixing bacteria than older, less domesticated varieties [119]. Thus, plants adapted to "optimal" agricultural conditions might be less able to form effective symbioses than their ancestors, further excluding certain microorganisms from these human-altered systems.

Furthermore, planting the same cultivars across ecologically distinct regions with different indigenous soil communities is likely to reduce the native soil biodiversity. Part of this reduction in native soil communities could be due to the cultivars preferentially promoting interactions with the microbial taxa with which they were bred, instead of with native microbes [114, 120]. Planting varieties that are not adapted to the local microflora might not only reduce the efficiency of crop growth but also promote the homogenization of soil communities across regions, as unique members of the native communities are displaced by microbes promoted by the chosen cultivar. Coupled with the fact that most crops will be planted in genetic (as well as ontogenetic) monocultures, modern agriculture and breeding practices likely place a strain on soil biodiversity [93, 114].

4.2 Considerations for studying plant-microbe interactions in human-altered habitats

The effects of human disturbances on natural systems can be difficult to study, as many human activities cause drastic and complex changes to ecosystems, alter the systems in multiple ways, and are often repeated over time [121]. However, such human-influenced systems also offer opportunities to study the mechanistic and functional aspects of plant-microbial interactions and their effects on ecological dynamics, since they frequently represent a mosaic of different degrees of disturbance. This section covers approaches and considerations for the design of studies that assess the impacts of human disturbances on habitats, with a particular focus on examining the effects of microbial community disturbance on plant performance and functional traits. To demonstrate issues that may arise during experimental planning, we discuss the design of a hypothetical project (illustrated in Fig. 4.1) that aims to investigate the impacts of agricultural management, and particularly tillage, on soil microbial communities and plant traits with particular attention to effects on plant growth, herbivore resistance, and metabolic phenotypes. Using this example, we discuss two general approaches to studying the effects of disturbances: (1) surveys across artificial or natural gradients of disturbance (Section 2.1) and (2) targeted manipulative experiments (Section 2.2).

4.2.1 Surveys along artificial or natural gradients of disturbance

A major advantage of studying human-altered habitats, versus those that were disturbed by other species or natural events, is that we are more likely to be able to realistically recreate the disturbance and/or identify disturbances from the same types of activities. For many common types of disturbed systems, such as agricultural fields, potential study sites are not only plentiful and relatively accessible but also relatively simple to create. One common approach for studying the effects of disturbances on an ecosystem is to compare habitats over a gradient of that disturbance. This may involve defining undisturbed *control* sites that are approximately equivalent to the sites of interest, but in relative *pre-disturbance* states, sites that are/were *actively disturbed* to different degrees, and/or in various stages of *post-disturbance*. To study sites that were disturbed to different degrees in our agricultural disturbance example, we could compare fallow fields that had been previously used for agriculture for varying numbers of years. Studying sites in different stages of post-disturbance could involve comparing agricultural fields in different stages of oldfield (post-agricultural) succession, which is illustrated in Fig. 4.1. In order to compare the effects of the disturbance on a plant across such a gradient, we selected a dominant species that is both an early colonizer of fallow agricultural fields and one that persists through many stages of oldfield succession, tall goldenrod, *Solidago altissima* (Asteraceae) [122], as a focal, indicator plant.

There are advantages and disadvantages to performing surveys across disturbance gradients compared to more targeted experimental approaches. The main advantage of gradients is that they allow studies in real “natural” environments in which a disturbance has actually occurred. This is particularly important for understanding the effects that the disturbance might have on ecological interactions in the disturbed community, such as interactions between the plants with animals, microbes, and other plants, and may inform which traits warrant further study in targeted manipulative experiments. The main disadvantage in studying gradients of disturbances, especially natural gradients, is that one will be able to assess correlations only between the disturbances and observed microbial and plant communities and traits, as the

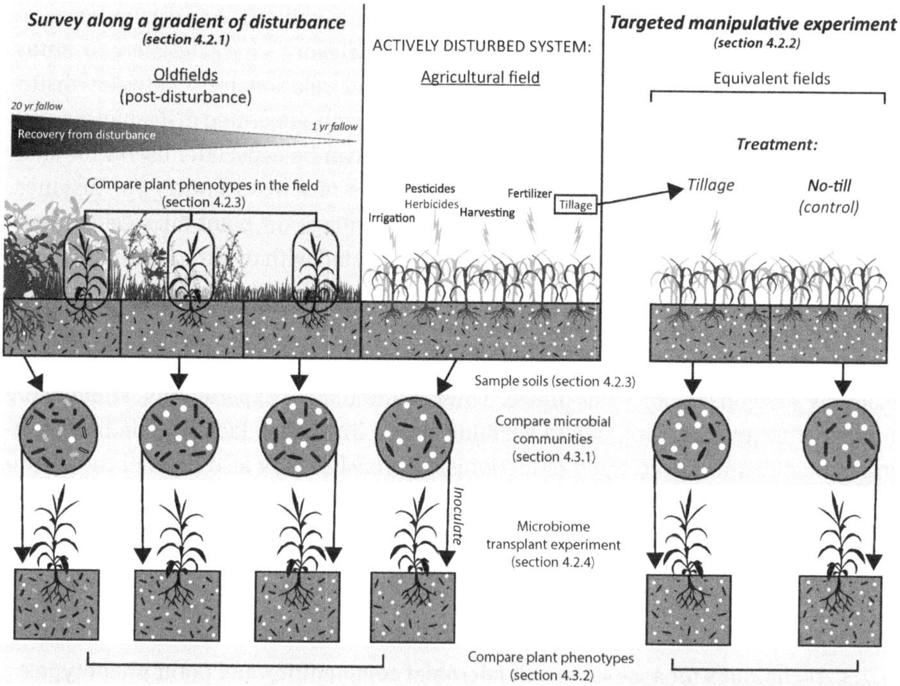

Fig. 4.1: Summary of approaches and methods for studying the effects of disturbances on plant microbe interactions using an agricultural disturbance example. The effects of agriculture can be assessed via field surveys across a gradient of disturbance (leftmost panel; section 4.2.1) as well as more mechanistically through manipulative experiments (right panel; section 4.2.2). This example illustrates how a single mechanism of disturbance, tillage, can be specifically targeted in a manipulative study. Comprehensive studies of disturbances on plant-microbe interactions may involve sampling (Section 4.2.3) and comparison (Section 4.3.1) of microbial communities and plant phenotypes (Section 4.3.2) in the field (Section 4.2.3), as well as the examination of the effects of the disturbed microbial communities on plant phenotypes (Sections 4.2.4 and 4.3.2).

disturbance is likely not the only factor to vary between sites. In studying complex disturbances, such as agricultural cultivation, that alter the environment through multiple activities (e.g. tillage, fertilizer, and herbicide application) and mechanisms (Tab. 4.1), it may also be difficult to isolate the effects of a particular activity on plant-microbe relationships without taking a more experimental approach.

4.2.2 Targeted manipulative experiments

Experiments that manipulate a disturbance or a particular part of a disturbance, while controlling for other potentially confounding factors, allow one to specifically examine the consequences of that disturbance on a habitat. Whereas surveys are useful for identifying patterns and ecologically relevant characteristics to study, manipulative studies allow examination of causal relationships between disturbances and such consequences. For complicated human-generated disturbances such as agriculture, targeted manipulative experiments can be especially useful for identifying the consequences of individual components of a disturbance. For instance, if one aims to specifically examine the effects of tillage on plant-microbe relationships, comparing otherwise equivalent fields cultivated with or without tillage would be an appropriate experiment (Fig. 4.1). Of course, there are also factors such as weed control practices that might typically vary with tillage practices, but adding additional treatments without the proper controls would prevent a researcher from isolating the consequences of tillage. Thus, manipulative experiments, while being reproducible and allowing one to examine causal links, may lack realism in recreating human disturbances. Such reductionist approaches may also prevent the identification of important interactions between multiple mechanisms of disturbance, as well as observations of how the disturbance affects the interactions of plants and/or microbes with their natural environment.

4.2.3 Techniques for assessing soil microbial communities and plant phenotypes in the field

Representative and consistent sampling of microbes and plants can be difficult to achieve in a field setting but is a critical part of assessing the effects of disturbances on plant-microbe interactions. In the following sections, we discuss methods for sampling soil microbial communities and profiling plant phenotypes, which are applicable to both field survey and manipulative experimental studies. Standardization of sampling techniques is particularly important when making comparisons between such studies. Selecting specific plant species of interest on which to examine disturbance-based phenotypic changes can improve standardization and sampling consistency, as plant rhizosphere microbiomes and microbially mediated effects can be plant species specific [106, 123].

Sampling soil microbial communities in the field

Soil microbial communities can be highly variable, even within sites with homogenous aboveground plant communities, such as monoculture crop fields [124]. Consistent sampling is therefore important to avoid adding additional technical variance. The sampling scheme for surveying soil microbial communities at a disturbed site will depend on the disturbance. In determining the number and depth of samples to take, the magnitude (i.e. intensity), sequence (i.e. frequency, duration, and time since last disturbance), and heterogeneity (i.e. how evenly the disturbance affects the environment) of the disturbance should be considered. For example, in assessing the magnitude of agricultural tillage in a plot's history, we might consider what methods of tillage were used and how deeply the fields were tilled. Deeper and more intense tillage (e.g. using a rotary hoe) is likely to disturb the soil at greater depths [125, 126]. It is also helpful to consider the length of time since the field was last tilled, as well as the frequency of tillage events and the length of time that the fields have been cultivated, as the effects of disturbance on biological communities can be cumulative [127]. In terms of assessing the heterogeneity of the disturbance, if the tillage was evenly applied to a field (as would be expected with a mechanical plow), the disturbance would be expected to be experienced relatively evenly throughout the affected soil, whereas fields tilled by hand may be more heterogeneous. The time that a plot has been in a post-disturbance state may also affect heterogeneity, as the effects of the disturbance become diffuse, and factors other than the disturbance may increase in relative importance in shaping the habitat. In our field survey example, we would expect heterogeneity to increase with time since active disturbance, as the plots undergo community succession and factors such as competition and herbivory, which were previously controlled under agricultural conditions, may increasingly shape community composition [128, 129]. For unintentional disturbances, such as pollutants, understanding the heterogeneity of the disturbance may be of greater importance than those, such as agricultural amendments, that were applied with deliberate attention to evenness.

In determining the depth of soil samples, it may also be helpful to consider the root systems of the plant(s) of interest. Since the relevant fraction of the soil microbial community that affects plant traits includes those that colonize and interact with the roots, sampling at a depth at which the majority of the active roots are located will target the organisms that are likely to interact with the plant (rhizosphere soil collection is discussed in Section 3.1). Taking soil microbial community samples from the rhizosphere of individuals of a single plant species growing across different sites, in our case *S. altissima*, is one way to reduce variability in sampling, as the communities selected by plants in the rhizosphere are often distinct from the bulk soil and may even be plant species specific [130–132].

To understand the functional impacts of changes in the composition of a plant-associated microbial community, studies cannot remain restricted to the soil or even the rhizosphere community but need to include the derived endophyte communities that reside inside the plant. Many endophytic microbes are known to affect plant growth, as well as other phenotypes such as pathogen resistance [133, 134]. While endophytes may colonize plants through many routes, including vertical and horizontal transmission from other plants, it is thought that most endophytes are recruited from the rhizosphere, and substantial overlap has been observed between rhizosphere and endosphere communities [130, 131]. Thus, community assembly inside the plant can also largely be affected by soil conditions [133]. Because soil is not the only source of colonization, it is also possible that some, but not all, of the endophytes will be found in the rhizosphere pool. A microbe found in the endosphere is intimately associated with the plant through physical contact, as opposed to one that is simply located *near* the plant in the rhizosphere, which may mean that it is more likely to be directly interacting with the plant in a meaningful way. This makes the study of endophytes especially valuable. Surveying endophytic communities typically involves surface sterilization of the plant tissue to eliminate topical microbes and then extraction of endophytes from macerated tissue [133].

Profiling plants in the field

As plants are the main primary producers in most terrestrial ecosystems, understanding the effects of soil microbes on plant growth and phenotypes may be one of the most ecologically important measures of functional shifts in soil microbial communities. Observing phenotypic differences in the field is an important step that can be a source of motivation for most studies of human disturbance and indicate whether or not more manipulative or mechanistic studies are worth pursuing. Making careful observations in the field may also inform which traits would be interesting metrics for a particular study (characterizing plant phenotypes is discussed in greater detail in Section 3.2). For example, observing differences in the density of insect herbivores across a disturbance gradient may indicate that studying insect defense-related traits, such as secondary metabolite production in leaves, could be of interest, although care should be taken to avoid spurious conclusions based on peculiarities of a specific site or experimental setup [135].

Potential traits to measure from plants in the field:

1. Size (e.g. height, biomass, and number of leaves)
2. Fitness (sexual or asexual reproduction)
3. Herbivore density and damage
4. Pathogen infection
5. Pollinator visitation rate
6. Leaf nutritional quality (e.g. chlorophyll content, C/N)
7. Photosynthetic rates
8. Herbivore defense traits
9. Plant metabolites (e.g. leaf metabolites, total VOC metabolome)

Studying plant phenotypes in the field is critical to forming hypotheses about the effects of a disturbance on plant communities, as plant phenotypes, including those involving plant-microbe interactions such as pathogen resistance [136] and mycorrhizal symbioses [137], that are observed in the laboratory or greenhouse are not always observed to an equivalent degree in the field. Plant phenotypes in field studies have also been shown to differ from those observed under controlled conditions [138]. However, the complexity of field ecosystems makes it difficult to establish causal links between the disturbance and changes in relevant plant traits. Observing differences in the field is also likely to reveal traits that may have ecological relevance and indicate whether more controlled experiments are needed to identify the mechanisms through which a particular disturbance affects plant phenotypes. Disturbances can both directly and indirectly affect plant phenotypes, which makes manipulative experiments important for disentangling potentially confounding effects of different disturbance-affected factors on plant traits. Additionally, differences observed in plant traits in the field might not all be due to phenotypic plasticity. Long-term disturbances, for example, may also have had microevolutionary effects on plant populations that explain the observed differences in phenotypes [139, 140].

4.2.4 Examining the effects of soil microbial shifts through microbiome transplant experiments

Observing concurrent changes in soil microbial communities and plant traits after a disturbance does not necessarily imply that shifts in plant phenotypes are microbially mediated. Many anthropogenic disturbances can affect plants directly, as well as indirectly through mechanisms other than microbial interaction. For example, the addition of P fertilizer directly promotes plant growth, but also reduces plant-mycorrhizal symbioses, which can affect the plant in other ways, such as by decreasing the ability of the plant to access micronutrients in the soil [117]. The extent to which plants directly engage with soil microorganisms is also likely to vary based on the specific plant species and environmental conditions, as plants vary in the amount and form of carbon and other nutrients that they allocate to belowground communities through exudation [141–143]. It is also possible that a disturbance could separately affect both plants and microbial communities with no direct interaction. It is therefore difficult in the field to differentiate the effects of microbial communities on plants from the effects of other environmental factors. For example, in our case study, agricultural tillage can affect plants by changing the soil texture and nutrient content, in addition to affecting the microbial community with which its roots interact [61].

While observing changes to microbial communities and plant traits at disturbed sites is important for understanding the integrated effect of the disturbance *in situ*, controlled manipulative experiments are crucial for examining the effects of the disturbance on plants and microorganisms and particularly microbially mediated plant traits. It may be important in some studies to examine the effects of a disturbance on microbes and plants separately, as well as in combination. For example, in soils contaminated with petroleum, bacteria were only found to increase expression of hydrocarbon-degrading genes in association with a plant [144], which emphasized the importance of the interaction between plants and microbes for bioremediation. However, the focus of this section is on examining how disturbances to soil microbial communities affect the expression of plant phenotypes.

Designing soil microbiome transplant experiments

Microbiome transfer experiments are one approach for isolating the effects of soil microbial communities on plant traits [99, 100], and can help to examine the indirect effects on plants of imposing extreme environmental conditions on microbes. A soil microbiome transfer attempts to transplant a whole microbiome from a source soil to a sterilized recipient soil medium. Choice of plants, the recipient medium, and the method of transfer should be carefully considered in the design of the experimental methods.

Selection of plants: In selecting indicator plants for studying microbe-altered plant phenotypes, several considerations should be made to control for other sources of variation that might affect microbial colonization. The microbial community that assembles around a plant can be plant species specific, even within closely related species [132]. Host species have also been shown to differentially affect which rhizosphere taxa are active [145]; thus, choosing which plant species to use as an indicator for microbe-altered plant phenotypes is not arbitrary. For the greatest relevance, ideally, one would use a plant species that is likely to be affected by the disturbance of interest. In the case of our tillage example, *S. altissima* was chosen since it is found ubiquitously in the surrounding area and colonizes previously tilled fields, but any other oldfield species present at the site would also be relevant. Distinct rhizosphere microbiomes have been observed even among species of the genus *Solidago* [132], so it is important to ensure that comparisons are made between plants of the same species and even subspecies and genotype. Evidence for plant genotypic effects on rhizosphere colonization have been observed in a range of species, including maize [146] and *Arabidopsis thaliana* [147]. Genotypic effects on phyllosphere colonization have also been observed in the wild mustard, *Boechera stricta* [148].

If choosing plants from the field and indicator plants, particularly at the site of the disturbance of interest, it is important to consider how the disturbance may have been an agent of natural selection on the plant population. Rapid microevolutionary changes in *S. altissima*, for example, have been observed in plant populations after 12 years of pesticide application. These changes affected the production of allelopathic polyacetylene compounds in the roots, which could potentially affect rhizosphere interactions [140]. In order to tease out potential confounding microevolutionary effects, using plant genotypes collected both from the disturbed area and an adjacent undisturbed area is desirable.

The ease with which plant material can be sterilized and cultivated under greenhouse conditions is also a practical consideration. Seeds, which are commonly surface sterilized using bleach or ethanol, can perhaps be sterilized most effectively than plant tissue, although the seeds of many plant species host a variety of endophytic fungi and bacteria that are unlikely to be removed with surface treatment [149, 150]. Similarly, clonally propagated cuttings, which have the advantage of genetic sameness, are more difficult to sterilize, as surface sterilization is unlikely to eliminate endophytic microbes. *Solidago* is self-incompatible, and therefore, clonal propagation is the only option to replicate genotypes, so the importance of reducing genetic variability must be weighed against the importance of removing a preexisting endosphere microbiome. While surface sterilization did not eliminate microbes inside of the much larger seeds of a confamilial plant, sunflower (*Helianthus annus*), seed endophytes had little effect on the establishment of the root and rhizosphere microbiome [150]. The difficulty and speed of growing plants from seeds versus cuttings must also be considered for the species of interest, as the soil microbial community in a pot may diverge from the original inoculated community over time as the transplanted microbiome establishes and undergoes succession. Seeds with lengthy germination times and slow growing seedlings, for example, might encounter a distinct microbial community if they start producing roots only after several weeks, compared to a cutting that began growing shortly after inoculation. Spiking seeds or cuttings with inoculant at the time of planting may increase the chance of inoculation with the microbial community of interest.

Selection of a plant growth medium: The recipient soil medium, or the medium in which the plants will be cultivated, can affect both plant growth and the microbial community that assembles in the microcosm. Some abiotic soil properties that should be considered for their effects on plant growth, particularly when cultivating plants in an artificial setting, include nutrient availability and drainage. Transferring a soil microbiome to a novel environment, with different physical and chemical properties than its source, will undoubtedly exclude some and promote other specific taxa, altering the community composition relative to the source. Soil properties might also affect the behavior of microbes; for example, arbuscular mycorrhizal fungi were found to be more active, forming greater numbers of arbuscules, in their native soil compared to foreign soils [94]. Some soil properties that are likely to affect microbial community assembly, such as pH, nutrient availability, and soil texture, are discussed in detail in Section 1.1.

Standard commercial potting media (typically sphagnum peat moss or coir based) is widely available and generally supports the growth of most plants in a greenhouse environment. However, not containing a mineral component (i.e. sand, silt, and clay), this "soilless medium" is quite different from field soils and may not support a similar microbiome. Soil texture has been found to be associated with microbial community composition and diversity [151, 152], and thus, transplanting a microbial community from a natural soil to a medium with dissimilar physical properties is likely to selectively promote the growth of a subset of the community, resulting in a composition distinct from that of the donor soil. Using sterilized parent field soil as a recipient medium is one way to emulate the physical properties of the microbes' native habitat, although common sterilization procedures, such as steam sterilization and gamma-irradiation, can alter soil properties such as texture and organic matter [153]. Lastly, the presence or absence of particular soil compounds in the recipient media that inhibit (e.g. allelopathic chemicals) or promote microbial growth could affect the composition of the microbial community that assembles [154].

Most importantly, the medium must be conducive to growing the plant of interest, even if this means making compromises with media that support assembly of the microbiome most representative of that found in the field. In our case study, *S. altissima* can be propagated vegetatively via rhizome cuttings and the plants are cloned this way in many studies to create isogenic plant material (as a self-incompatible species, plants grown from seeds will not be isogenic) [155]. It may be difficult to cultivate *S. altissima* plants in a greenhouse setting from these cuttings, which require extensive moisture, in a heavy clay field soil that is prone to compaction. A compromise may involve mixing in perlite or calcined clay (e.g. Turface) to the sterilized field soil to improve aeration and drainage. It is also important to consider whether or not the initial nutrient content of the media will be sufficient to support the growth of the selected plant to the developmental stage the plant requires, as adding fertilizer throughout the experiment can also alter the microbial community. Thus, ideally, several media preparations and fertilizer regimes should be tested for both microbiome transferability and plant cultivability.

Selection of microbiome transfer method: In addition to the choice of recipient medium, the method of microbiome transfer can also strongly affect the microbiome that assembles in a microcosm. In selecting transfer methods, the importance of limiting the concurrent transfer of abiotic factors such as soil compounds alongside the soil microbes, as well as maintaining a representative community, should be considered. The simplest, and most representative, method of soil microbiome transfer seems to be direct inoculation of the recipient with field soil containing the microbiome of interest, and this has been widely used [99, 100, 156]. This method is free from the bias of selection through culturing or filtering but, in directly transferring parent soil into the microcosm, has the risk of transferring potentially influential soil compounds that may have confounding effects on plant traits. This risk can be minimized by transferring only a small volume of inoculant soil, although inoculating with increasingly dilute concentrations decreases the resemblance of the transferred microbial community to that of the source soil, as well as community diversity [157, 158].

Using a soil wash microbiome transfer technique can further minimize the risk of transferring soil compounds. In soil washes, the microbial fraction of soil is extracted in a dilute aqueous salt solution and used for inoculation. This can be achieved by shaking the donor soil in solution, removing soil particles via filtration (with a large enough pore size to not exclude microbes), pelleting the microorganisms via centrifugation, and then resuspending the microbial fraction in a sterilized salt solution, and hence minimizing the transfer of abiotic soil factors [159]. However, soil washes have been found to decrease the diversity of transferred bacterial and fungal communities, resulting in communities that were less similar to the donor soil compared to direct soil inoculations of equivalent soil volume [158]. Thus, minimizing the transfer of soil compounds comes at a cost. A third, even more selective approach, is to inoculate with the cultivable fraction of the microbial community. While having the advantage of reproducibility and the potential of scaling up, taking a cultivation-based approach alters the composition of the transferred microbial community by promoting those that thrive under the culture conditions and excluding those that do not. Such an approach was used to inoculate soil microcosms that were treated with crude oil, resulting in less biodegradation than was observed in soil inoculated through direct soil transfer [160]. Despite alterations in community composition, soil microbiomes transferred via cultured inoculants have been shown to affect plant phenotypes in a similar manner to their direct soil inoculated counterparts [161].

It is also important to note that the process of performing soil microbiome transplants causes disturbances to microbial inoculant. Excavating and homogenizing the soil inoculant prior to the transfer cause physical disturbances to the microbial community, such as breaking fungal hyphae and increasing aerobic conditions, which could alter the viable microbial community (see Section 1.4 for a discussion of physical soil disturbances). Processes such as filtering, centrifugation, and culturing are likely to cause further disturbances to the microbial community. During transplantation, microbes are also introduced into a vastly different competitive environment than their native soil, as the relatively tiny community of microbes in the inoculant is released into a large volume of sterilized soil. This may provide an opportunity for species previously kept at low abundances due to competition to become more prominent [58] or may favor early successional species. Little is known about how long it takes for a microbiome transplant to colonize a new environment or how long it takes for the community to stabilize, and it is unclear whether or not allowing time for stabilization prior to adding plants to the mesocosms would ultimately influence assembly of the microbial community. Studies addressing these issues are urgently needed, because they would provide the basis for the creation of collection and processing standards in soil microbiome manipulative experiments.

In conclusion, it is important to consider the potential effects of transferring soil compounds alongside microbes, as well as the effects of disturbances and selection on microbiome integrity, in selecting a microbiome transfer approach. In our example of studying the effects of tillage, the risk of transferring potent soil compounds is likely to be relatively low, and therefore, a direct soil inoculation, which transfers a community with minimal selection bias, will probably be the best choice. On the other hand, for a study investigating the effects of a potent compound, such as a pollutant, that is stable in the soil, a soil wash or cultured inoculant has the benefit of transferring the microbial community (albeit one that has undergone some selection based on the transfer method) without concomitantly transferring the compound. Best practice would be to compare several potential transfer methods at multiple inoculation concentrations in order to select a method that effectively transfers a representative microbiome or desired groups of microbial taxa.

4.3 Analysis of plants and associated microbiomes

In order to assess the microbially mediated effects of human disturbances on plants, it is important to establish (1) that the disturbances to microbial communities are reflected in the microbial communities that associate with plants and (2) that these changes in the plant microbiome result in altered plant phenotypes. Methods for characterizing plant-associated microbiomes and profiling plant phenotypes are discussed in the following sections.

4.3.1 Profiling plant-associated microbiomes

Characterizing the communities of microbes that closely associate with plants is important for assessing the effects that alterations in microbial communities could be having on plant phenotypes. Microbes in the rhizosphere and endosphere are likely to be interacting with the plant and affect its metabolism and behavior. Due to its affordability and scalability, the compositions of microbial communities are most commonly analyzed using high-throughput amplicon sequencing to identify taxa (see Chapters 7 and 9). Specific studies will demand additional molecular/omics approaches, but we will not discuss those in detail here.

Given that plant-associated soil communities are highly heterogeneous in both time and space [115], it may be useful to consider using composite soil samples in order to increase the representativeness of samples while keeping the cost and labor associated with sample analysis manageable. For example, portions of the plant root systems at different developmental stages, such as the tips of the roots versus the base, are also likely to harbor functionally different communities of microbes, potentially as a result of differences in the release of root exudates [162]. Thus, in taking rhizosphere samples, soil collected from multiple portions of the roots of a plant could result in the collection of a more representative pool of microbes with which the plant's roots interact. Mature *S. altissima* plants can have large fibrous root systems, and therefore, combining soil samples from multiple sections of the roots will help capture the rhizosphere community. Extensive root sampling may be even more important for larger and more deep-rooted plants, such as trees, as their roots interact with microbial communities at a broader range of soil depths. Rates of root exudation have also been observed to decrease at greater depths [163], indicating that the communities of microbes supported by the rhizosphere may vary with root depth. In such cases, it may be useful to analyze soils collected at different depths separately or collect all samples at approximately the same depth.

Many methods for collecting the rhizosphere soil community (the community outside of, but immediately surrounding, the root) have been used [164]. The simplest procedure involves uprooting a plant and shaking off the adherent soil into a sterile container by hand, but more tightly attached soil can be separated from the root using a fine sterile brush [165] or dislodged by vibrating small pieces of root in a dry tube using a vortex [166]. Rhizosphere soil has also been collected in solution by washing or shaking off adherent soil in a buffered solution and potentially using centrifugation to isolate the microbial fraction from the soil particles [164].

4.3.2 Profiling plant phenotypes

In order to assess the effects of disturbances and disturbed soil microbial communities on plant traits, the phenotypes of plants must be characterized and compared to control (undisturbed) plants or across a gradient of disturbance. There are nearly endless traits to potentially assess, and it is recommended that preliminary field observations be used to inform which phenotypes might be of greatest interest to compare. However, with respect to species interactions and the overall effects on community and ecosystem dynamics, measures of plant reproductive success (the effect of the altered environment on the organism) and the plants secondary metabolism (the effect that the altered plant phenotype has on interactions with the environment) are most likely candidates. Thus, in this section, the characterization of plant phenotypes with respect to growth and reproduction, chemistry, and ecological interactions is discussed, along with potential complications of studying these traits in disturbed systems.

Growth and reproduction

Some of the most obvious and ecologically pertinent effects of a disturbance on plant communities may be those that affect plant growth and reproduction. Many disturbances, such as heavy metal pollution [167] and physical soil disturbances [168], reduce the growth and fitness of plants. On the other hand, some disturbances, such as fertilizer application, can promote plant growth. While many disturbances can directly affect plants, they may also influence plant growth and reproduction through altering plant-microbe relationships.

One of the simplest, but most important, plant traits to measure is size. The growth of plants is also known to be affected by soil microbes, including mycorrhizal fungi [94] and aptly-named "plant growth-promoting rhizobacteria" [95]. Assessing the size of plants is useful not only because it may indicate the vigor or a plant but also because it may be useful to standardize many other measurements (e.g. amount of herbivore damage) to the overall size of a plant. Depending on the species, size measurements can range from simple (e.g. biomass of an *Arabidopsis* plant) to complex (e.g. estimating the size of a large woody plant). In terms of our example, *S. altissima* is a large herb on which height measurements can be easily performed (typically less than 2 m in height). Mature plants can have many ramets and hundreds of leaves, making leaf counts in the field potentially time-consuming, although plants are easy to collect for later counts and dry for biomass measurements.

The reproductive output of a plant is often of interest as an indicator of the effect of the disturbance on plant fitness. Traits such as flower number, fruit production, seed production, and offspring viability are often used to assess sexual reproduction, while asexual reproduction can be assessed in many species as production of new ramets or rhizomes for clonal reproduction. In *S. altissima*, inflorescence size is probably the most practical measurement of sexual reproduction as the plants can produce thousands of individual flowers and produce thousands of tiny seeds. *Solidago altissima* also reproduces clonally via underground rhizomes, which can be excavated from the field [169]. It is important to note that sexual reproduction is often induced by stress in plants [170] and measuring reproductive structures is not a direct measure of the fitness of a plant. Particularly, the relationship between measurements of reproductive structures or output and plant fitness might change when an ecosystem is disturbed. For example, inflorescence size might not be as strong of a predictor of fitness in a system in which the disturbance limits the activity of pollinators [171].

Chemical phenotypes

Many ecological interactions between a plant and other organisms are mediated by plant chemistry [172, 173]. Thus, the chemical profiles of a plant can be important determinants of how a plant interacts with herbivores, other plants, microbes, and others. Some human disturbances, such as chemical pollutants, can directly alter the chemical profiles of plants, as contaminants are assimilated and metabolized by plants [174, 175]. Soil microbial communities have also been shown to affect the metabolic profiles of plants [106], and thus, disturbances to plant-associated microbial communities have the potential to not only alter a plant's chemistry, but also the plant's chemically-mediated interactions.

If the chemicals that mediate specific plant interactions of interest are known, one may want to focus on analyzing plant tissue for those compounds specifically. For example, *S. altissima* plants are known to produce the allelopathic compound dehydromatricaria ester in their roots, which inhibits the growth of neighboring plants [140]. Thus, in our agricultural disturbance example, to understand how disturbed microbial communities affect the competitiveness of *S. altissima* plants, we might examine whether soil microbial communities are affecting the amount of this particular allelopathic compound by analyzing concentrations in root tissue samples [140].

However, because the chemical compounds that mediate the interactions of interest are often unknown (or not all known), as well as which secondary metabolites will be affected by a disturbance, it can be useful to take an unbiased metabolomics approach in chemically phenotyping and comparing plants. While none of the current methods for extracting and analyzing snapshots of chemically diverse plant metabolomes are entirely comprehensive, untargeted metabolomics approaches can be very useful for characterizing plant responses and identifying potentially important compounds that may mediate plant interactions [176]. As many important plant interactions are mediated via volatile organic compounds (VOCs) [172, 173], characterizing and comparing the total VOC profiles of plants exposed to different disturbances or microbial communities is one example of a metabolomics approach to broadly assess the effects of treatments on plant metabolites. One way to assess the VOC metabolome of a plant is collect VOCs from the plant's headspace and analyze the collected compounds using GC-MS [177–179]. Similar to the composition of a microbial community, the metabolite profile of a plant can be characterized using various multivariate similarity measures, which can be used to compare the profiles across plants, for example, using ordinations such as nonmetric multidimensional scaling [178].

It may be important to consider not only the effects that a disturbance might have on the secondary metabolism of a plant but also the potential effects that it may have on the fate and perception of those chemical changes by other organisms. For example, in the case of VOC emissions, atmospheric pollutants such as ozone and nitrogen oxides may react with these chemicals, altering the composition and longevity of the plant VOC emissions and ultimately potentially affecting the plants' interactions organisms that may perceive its VOC signals [180, 181]. Furthermore, many human disturbances, such as agricultural activities [182], directly add VOCs to the environment. In addition to potentially affecting biological processes, chemical pollutants that volatilize can add "noise" to the air, diluting plant VOC emissions and consequently making with them more difficult to perceive [183]. Moreover, disturbances that alter airflow in a plant system, such as the removal of tall plants that may serve as a windbreak, may alter the fate of VOC plumes by changing their persistence and the distance that they travel [184].

Integrative measures of phenotype on plant interactions (bioassays)

Ecologists are often interested in changes in plant phenotypes in the context that they alter the plant's interactions with its community. In addition to bacteria and fungi, many plants have important interactions, either mutualistic or antagonistic, with herbivores and other plants. Studying the changes in a plant's community interactions that result from disturbing its associated microbial community allows one to assess phenotypic changes in a broader ecological context.

Plant-associated microbes have been shown to affect plant defense traits, thereby altering their interactions with herbivores [102, 103]. One method for quantitatively comparing the relative attractiveness of plants to herbivores is to assess feeding preference in a choice bioassay. In a typical choice test experiment, a herbivore is presented with a choice between multiple plants (or pieces of plant tissue) from different treatments, and after an allotted amount of time, the amount of tissue consumed of each plant is measured [155, 185]. Presumably, the herbivore will consume a greater amount of the plant that is less well defended. In our example, we could use a choice test assay to determine whether microbial communities disturbed by agriculture affect the attractiveness of *S. altissima* plants to an insect herbivore, such as the goldenrod leaf beetle, *Trirhabda virgata*, by comparing the beetle's preference for a plant grown with a disturbed microbiome versus an undisturbed control microbiome. Another common bioassay used to assess plant resistance to herbivores involves measuring amounts of consumption and herbivore biomass assimilation when reared on different plants, as herbivores will presumably gain biomass less efficiently when plants are more strongly defended [185, 186]. For example, if we found that *T. virgata* larvae gained more weight for every square millimeter of tissue eaten when feeding on *S. altissima* plants with disturbed microbiomes compared to their counterparts with undisturbed microbiomes, it would indicate that those plants were less herbivore resistant.

In addition to interactions with herbivores, plants also antagonistically interact with other plants. Soil microbial communities have also been shown to affect competitive plant-plant interactions both by degrading allelopathic compounds [154, 187], as well as altering the allelopathic chemistry of plants [106]. Endophytic microbes can also affect a plant's allelopathic potential [188]. Simple allelopathy bioassays can be performed by assessing the effects of a plant tissue extract on seed germination rate [106, 188] or by assessing the growth rates of plants grown in competition with an indicator plant [140].

4.4 Concluding remarks

The drastic environmental changes brought about by human activities can generate extreme conditions for soil microorganisms, which ultimately impact the greater communities with which they interact. Here, we have focused on the effects that disturbances to soil microbes have on plant-centered ecosystems, such as agricultural fields, as these systems are both highly disturbed and of great importance to humans. Studying the integrated effects of such disturbed habitats is a challenge, and we suggest that combined profiling of microbial communities and plant phenotypes, using both surveys and manipulative studies, will help us understand how human activities ecologically impact the environment. While this chapter has been primarily written from a plant-centric point of view, as plants will make up the bulk of primary producers in most terrestrial ecosystems disturbed by humans, the general framework for approaching such studies that we present could potentially be applied to non-plant systems. As we continue to disturb the soil through our daily activities, understanding the functional impacts of our actions will be critical for making decisions with regard to environmental management.

References

[1] Tilman D, Lehman C. Human-caused environmental change : Impacts on plant diversity and evolution. Proc Natl Acad Sci. 1987;98:5433–40.
[2] Western D. Human-modified ecosystems and future evolution. Proc Natl Acad Sci. 2001;98:5458–65.
[3] Grime JP. Evidence for the existence of three primary strategies in plants and its relevance for ecological and evolutionary theory. Am Nat. 1977;111:1169–94.
[4] Rothschild LJ, Mancinelli RL. Life in extreme environments. Nature. 2007;408:1092–101.

[5] Bassil NM, Bryan N, Lloyd JR. Microbial degradation of isosaccharinic acid at high pH. ISME J. 2014;9:310–20.
[6] Baker BJ, Banfield JF. Microbial communities in acid mine drainage. FEMS Microbiol Ecol. 2003;44:139–52.
[7] Violle C, Pu Z, Jiang L. Experimental demonstration of the importance of competition under disturbance. Proc Natl Acad Sci. 2010;107:12925–9.
[8] Wubs ERJ, van der Putten WH, Bosch M, Bezemer TM. Soil inoculation steers restoration of terrestrial ecosystems. Nat Plants. 2016;2:16107.
[9] Friesen ML. Microbially mediated plant functional traits. Annu Rev Ecol Evol Syst. 2013;1:87–102.
[10] Fox JE, Gulledge J, Engelhaupt E, Burow ME, McLachlan JA. Pesticides reduce symbiotic efficiency of nitrogen-fixing rhizobia and host plants. Proc Natl Acad Sci U S A. 2007;104:10282–7.
[11] Zhu Q, Riley WJ, Tang J. A new theory of plant-microbe nutrient competition resolves inconsistencies between observations and model predictions. Ecol Appl. 2017;27:875–86.
[12] Ghaly A, Ramakrishnan V. Nitrogen sources and cycling in the ecosystem and its role in air, water and soil pollution: a critical review. J Pollut Eff Control. 2015;3:1–26.
[13] Sharma SB, Sayyed RZ, Trivedi MH, Gobi TA. Phosphate solubilizing microbes: sustainable approach for managing phosphorus deficiency in agricultural soils. Springerplus. 2013;2:587.
[14] Nannipieri P, Giagnoni L, Landi L, Renella G. Role of phosphatase enzymes in soil. In: Bünemann E, Oberson A, Frossard E, editors. Phosphorus in action: biological processes in soil phosphorus cycling. Berlin, Heidelberg: Springer Berlin Heidelberg; 2011. p. 215–43.
[15] Jakobsen I, Leggett ME, Richardson AE. Rhizosphere microorganisms and plant phosphorus uptake. In: Sims J, Sharpley AN, editors. Phosphorus: agriculture and the environment, agron. Monogr. 46. Madison, WI: ASA, CSSA, and SSSA; 2005. p. 437–94.
[16] Jansa J, Finlay R, Wallander H, Smith FA, Smith SE. Role of mycorrhizal symbioses in phosphorus cycling. In: Bünemann E, Oberson A, Frossard E, editors. Phosphorus in action: biological processes in soil phosphorus cycling. Berlin, Heidelberg: Springer; 2011: 137–68.
[17] Liu Q, Qiao N, Xu X, et al. Nitrogen acquisition by plants and microorganisms in a temperate grassland. Sci Rep. 2016;6:22642.
[18] Galloway JN, Dentener FJ, Capone DG, et al. Nitrogen cycles: past, present, and future. Biogeochemistry. 2004;70:153–226.
[19] Eo J, Park K-C. Long-term effects of imbalanced fertilization on the composition and diversity of soil bacterial community. Agric Ecosyst Environ. 2016;231:176–82.
[20] Li F, Chen L, Zhang J, Yin J, Huang S. Bacterial community structure after long-term organic and inorganic fertilization reveals important associations between soil nutrients and specific taxa involved in nutrient transformations. Front Microbiol. 2017;8:1–12.
[21] Pan Y, Cassman N, de Hollander M, et al. Impact of long-term N, P, K, and NPK fertilization on the composition and potential functions of the bacterial community in grassland soil. FEMS Microbiol Ecol. 2014;90:195–205.
[22] Hartmann M, Frey B, Mayer J, Mader P. Distinct soil microbial diversity under long-term organic and conventional farming. ISME J. 2015;9:1177–94.
[23] Wang J, Song Y, Ma T, et al. Impacts of inorganic and organic fertilization treatments on bacterial and fungal communities in a paddy soil. Appl Soil Ecol. 2017;112:42–50.
[24] Leidi EO, Rodriguez-Navarro DN. Nitrogen and phosphorus availability limit N xation in bean. New Phytol. 2000;147:337–46.
[25] Zheng B, Hao X, Ding K, Zhou G, Chen Q. Long-term nitrogen fertilization decreased the abundance of inorganic phosphate solubilizing bacteria in an alkaline soil. Sci Rep. 2017;7:e42284.

[26] Heckman JR, Sims JT, Beegle DB, et al. Nutrient removal by corn grain harvest. Agron J. 2003;95:587–91.
[27] Tan Z, Lal R. Global soil nutrient depletion and yield reduction. J Sustain Agric. 2005;26:123–46.
[28] Ikeda S, Sasaki K, Okubo T, et al. Low nitrogen fertilization adapts rice root microbiome to low nutrient environment by changing biogeochemical functions. Microbes Env. 2014;29;50–9.
[29] Leff JW et al. Consistent responses of soil microbial communities to elevated nutrient inputs in grasslands across the globe. P Natl Acad Sci USA. 2015; 112: 10967–10972.
[30] Riley D, Barber Sa. Effect of ammonium and nitrate fertilization on phosphorus uptake as related to root induced pH changes at the root soil interface. Soil Sci Soc Am Proc. 1971;35:300–6.
[31] Likens GE, Driscoll CT, Buso DC. Long-term effects of acid rain: response and recovery of a forest ecosystem. Sci. 1996;272:244–6.
[32] Gray NF. Environmental impact and remediation of acid mine drainage: a management problem. Environ Geol. 1997;30:62–71.
[33] Fierer N, Jackson RB. The diversity and biogeography of soil bacterial communities. Proc Natl Acad Sci. 2006;103:626–31.
[34] Rousk J, Baath E, Brookes PC, et al. Soil bacterial and fungal communities across a pH gradient in an arable soil. ISME J. 2010;4:1340–51.
[35] Lauber CL, Hamady M, Knight R, Fierer N. Pyrosequencing-based assessment of soil pH as a predictor of soil bacterial community structure at the continental scale. Appl Environ Microbiol. 2009;75:5111–20.
[36] Marusenko Y, Herckes P, Hall SJ. Distribution of polycyclic aromatic hydrocarbons in soils of an arid urban ecosystem. Water Air Soil Pollut. 2011;219:473–87.
[37] Kampichler C. Influence of heavy metals on the functional diversity of soil microbial communities influence of heavy metals on the functional diversity of soil microbial communities. Biol Fertil Soils. 1996;23:299–306.
[38] Xie Y, Fan J, Zhu W, et al. Effect of heavy metals pollution on soil microbial diversity and Bermuda grass genetic variation. Front Plant Sci. 2016;7:1–12.
[39] Stazi SR, Moscatelli MC, Papp R, et al. A Multi-biological assay approach to assess microbial diversity in arsenic (As) contaminated soils. Geomicrobiol J. 2017;34:183–92.
[40] Bazzaz FA, Carlson RW, Rolfe GL. The effect of heavy metals on plants: part I. inhibition of gas exchange in sunflower by Pb, Cd, Ni and Tl. Environ Pollut. 1974;7:241–6.
[41] Perfus-Barbeoch L, Leonhardt N, Vavasseur A, Forestier C. Heavy metal toxicity: cadmium permeates through calcium channels and disturbs the plant water status. Plant J. 2002;32:539–48.
[42] Sikkema J, de Bont JA, Poolman B. Mechanisms of membrane toxicity of hydrocarbons. Microbiol Rev. 1995;59:201–22.
[43] Baker JM. The effects of oils on plants. Environ Pollut. 1970;1:27–44.
[44] Hildebrandt U, Regvar M, Bothe H. Arbuscular mycorrhiza and heavy metal tolerance. Phytochemistry. 2007;68:139–46.
[45] Yang J, Kloepper JW, Ryu CM. Rhizosphere bacteria help plants tolerate abiotic stress. Trends Plant Sci. 2009;14:1–4.
[46] Karst J, Gaster J, Wiley E, Landhäusser SM. Stress differentially causes roots of tree seedlings to exude carbon. Tree Physiol. 2016;37:1–11.
[47] Rajapaksha RMCP, Bååth E, Ba E. Metal toxicity affects fungal and bacterial activities in soil differently. Appl Environ Microbiol. 2004;70:2966–73.
[48] Aislabie J, Lloyd-Jones G. Bacterial degradation of pesticides. Aust J Soil Res. 1995;33:925–42.
[49] Riah W, Laval K, Laroche-Ajzenberg E, Mougin C, Latour X, Trinsoutrot-Gattin I. Effects of pesticides on soil enzymes: a review. Environ Chem Lett. 2014;12:257–73.
[50] Jacobsen CS, Hjelmsø MH. Agricultural soils, pesticides and microbial diversity. Curr Opin Biotechnol. 2014;27:15–20.

[51] Imfeld G, Vuilleumier S. Measuring the effects of pesticides on bacterial communities in soil: a critical review. Eur J Soil Biol. 2012;49:22–30.
[52] Dangi SR, Gerik JS, Tirado-Corbalá R, Ajwa H. Soil microbial community structure and target organisms under different fumigation treatments. Appl Environ Soil Sci. 2015;2015:1–8.
[53] Stapleton JJ, Quick J, Devay JE. Soil solarization: Effects on soil properties, crop fertilization and plant growth. Soil Biol Biochem. 1985;17:369–73.
[54] Rovira AD. Studies on soil fumigation-I, Effects on ammonium, nitrate and phosphate in soil and on the growth, nutrition and yield of wheat. Soil Biol Biochem. 1975;8:241–7.
[55] Ibekwe AM, Papiernik SK, Gan J, et al. Impact of fumigants on soil microbial communities. Appl Environ Microbiol. 2001;67:3245–57.
[56] Li J, Huang B, Wang Q, et al. Effect of fumigation with chloropicrin on soil bacterial communities and genes encoding key enzymes involved in nitrogen cycling. Environ Pollut. 2017;227:534–42.
[57] Gelsomino A, Cacco G. Compositional shifts of bacterial groups in a solarized and amended soil as determined by denaturing gradient gel electrophoresis. Soil Biol Biochem. 2006;38:91–102.
[58] Francioli D, Schulz E, Purahong W, Buscot F, Reitz T. Reinoculation elucidates mechanisms of bacterial community assembly in soil and reveals undetected microbes. Biol Fertil Soils. 2016;52:1073–83.
[59] Choi S, Song H, Tripathi BM, Kerfahi D, Kim H, Adams M. Effect of experimental soil disturbance and recovery on structure and function of soil community: a metagenomic and metagenetic approach. Sci Rep. 2017;7:1–15.
[60] Ridge EH, Theodorou C. The effect of soil fumigation on microbial recolonization and mycorrhizal infection. Soil Biol Biochem. 1972;4:295–305.
[61] Smith JL, Collins HP, Crump AR, Bailey VL. Management of soil biota and their processes. In: Paul EA, editor. Soil microbiology, ecology and biochemistry. Elsevier Inc. Burlington, MA, USA; 2014. p. 539–72.
[62] Fierer N, Strickland MS, Liptzin D, Bradford MA, Cleveland CC. Global patterns in belowground communities. Ecol Lett. 2009;12:1238–49.
[63] Wilhelm WW, Johnson JMF, Hatfield JL, Vorhees WB, Linden DR. Crop and soil productivity response to corn residue removal: a literature review. Agron J. 2004;96:1–17.
[64] Balesdent J, Chenu C, Balabane M. Relationship of soil organic matter dynamics to physical protection and tillage. Soil Tillage Res. 2000;53:215–30.
[65] Lal R. Tillage effects on soil degradation, soil resilience, soil quality, and sustainability. Soil Tillage Res. 1993;27:1–8.
[66] Tisdall JM, Oades JM. Organic matter and water-stable aggregates in soils. J Soil Sci. 1982;33:141–63.
[67] Brockett BFT, Prescott CE, Grayston SJ. Soil moisture is the major factor influencing microbial community structure and enzyme activities across seven biogeoclimatic zones in western Canada. Soil Biol Biochem. 2012;44:9–20.
[68] Linn DM, Doran JW. Aerobic and anaerobic microbial populations in no-till and plowed soils. Soil Sci Soc Am J. 1984;48:794.
[69] Ludermann H, Arth I, Liesack W. Spatial changes in the bacterial community structure along a vertical oxygen gradient in flooded paddy soil cores. Appl Env Microbiol. 2000;66:754–62.
[70] Lynch JM, Bragg E. Microorganisms and soil aggregate stability. Adv Soil Sci. 1985;2:133–71.
[71] Duchicela J, Sullivan TS, Bontti E, Bever JD. Soil aggregate stability increase is strongly related to fungal community succession along an abandoned agricultural field chronosequence in the Bolivian Altiplano. J Appl Ecol. 2013;50:1266–73.
[72] Hamza MA, Anderson WK. Soil compaction in cropping systems: a review of the nature, causes and possible solutions. Soil Tillage Res. 2005;82:121–45.
[73] Lei SA. Soil compaction from human trampling, biking, and off-road motor vehicle activity in a Blackbrush (*Coleogyne ramossisima*) shrubland. Western North Am Nat. 2004;64:125–30.

[74] Tanner CB, Mamaril CP. Pasture soil compaction by animal traffic. Agron J. 1959;51:329.
[75] Warner A, Mosse B. Independent spread of vesicular-arbuscular mycorrhizal fungi in soil. Trans Br Mycol Soc. 1980;74:407–46.
[76] Sylvia DM. Quantification of external hyphae of vesicular-arbuscular mycorrhizal fungi. Methods Microbiol. 1992;24:53–65.
[77] Kabir Z, O'Halloran IP, Fyles JW, Hamel C. Seasonal changes of arbuscular mycorrhizal fungi as affected by tillage practices and fertilization: hyphal density and mycorrhizal root colonization. Plant Soil. 1997;192:285–93.
[78] Bowles TM, Jackson LE, Loeher M, Cavagnaro TR. Ecological intensification and arbuscular mycorrhizas: a meta-analysis of tillage and cover crop effects. J Appl Ecol. 2016;54:1–9.
[79] Vos M, Wolf AB, Jennings SJ, Kowalchuk GA. Micro-scale determinants of bacterial diversity in soil. FEMS Microbiol Rev. 2013;37:936–54.
[80] Thies JE, Grossman JM. The soil habitat and soil ecology. In: Norman Uphoff, Ball AS, Fernandes E, et al. editors. Biological approaches to sustainable soil systems. Boca Raton, FL: CRC Press; 2006. p. 59–78.
[81] Trenbath BR. Intercropping for the management of pests and diseases. F Crop Res. 1993;34:381–405.
[82] Autrique A, Potts MJ. The influence of mixed cropping on the control of potato bacterial wilt (*Pseudomonas solanacearum*). Ann Appl Biol. 1987;111:125–33.
[83] Garrett KA, Mundt CC. Host diversity can reduce potato late blight severity for focal and general patterns of primary inoculum. Phytopathology. 2000;90:1307–12.
[84] Marshall DR. The advantages and hazards of genetic homogeneity. Ann New York Acad Sci. 1977;287:1–20.
[85] Mundt CC. Use of multiline cultivars and cultivar mixtures for disease management. Annu Rev Phytopathol. 2002;40:381–410.
[86] Traquair JA. Etiology and control of orchard replant problems: a review. Can J Plant Pathol. 1984;6:54–62.
[87] Yang J, Ruegger PM, McKenry MV, Becker JO, Borneman J. Correlations between root-associated microorganisms and peach replant disease symptoms in a California soil. PLoS One. 2012;7:e46420.
[88] Yim B, Winkelmann T, Ding GC, Smalla K. Different bacterial communities in heat and gamma irradiation treated replant disease soils revealed by 16S rRNA gene analysis – contribution to improved aboveground apple plant growth? Front Microbiol. 2015;6:1–12.
[89] Schnitzer SA, Klironomos JN, HilleRisLambers J, et al. Soil microbes drive the classic plant diversity-productivity pattern. Ecology. 2011;92:296–303.
[90] Maron JL, Marler M, Klironomos JN, Cleveland CC. Soil fungal pathogens and the relationship between plant diversity and productivity. Ecol Lett. 2011;14:36–41.
[91] Latz E, Eisenhauer N, Rall C, et al. Plant diversity improves protection against soil-borne pathogens by fostering antagonistic bacterial communities. J Ecol. 2012;100:597–604.
[92] Prober SM, Leff JW, Bates ST, et al. Plant diversity predicts beta but not alpha diversity of soil microbes across grasslands worldwide. Ecol Lett. 2015;18:85–95.
[93] Tiemann LK, Grandy AS, Atkinson EE, Marin-Spiotta E, Mcdaniel MD. Crop rotational diversity enhances belowground communities and functions in an agroecosystem. Ecol Lett. 2015;18:761–71.
[94] Johnson NC, Wilson GWT, Bowker MA, Wilson JA, Miller RM. Resource limitation is a driver of local adaptation in mycorrhizal symbioses. Proc Natl Acad Sci. 2010;107:2093–8.
[95] Compant S, Clément C, Sessitsch A. Plant growth-promoting bacteria in the rhizo- and endosphere of plants: their role, colonization, mechanisms involved and prospects for utilization. Soil Biol Biochem. 2010;42:669–78.
[96] Baas P, Bell C, Mancini LM, Lee MN, Conant RT, Wallenstein MD. Phosphorus mobilizing consortium Mammoth P^{TM} enhances plant growth. Peer J. 2016;4:e2121.

[97] Hijri M. Analysis of a large dataset of mycorrhiza inoculation field trials on potato shows highly significant increases in yield. Mycorrhiza. 2016;26:209–14.
[98] Verbruggen E, van der Heijden MGA, Rillig MC, Kiers ET. Mycorrhizal fungal establishment in agricultural soils: factors determining inoculation success. New Phytol. 2013;197:1104–9.
[99] Lau JA, Lennon JT. Rapid responses of soil microorganisms improve plant fitness in novel environments. Proc Natl Acad Sci. 2012;109:14058–62.
[100] Panke-Buisse K, Poole AC, Goodrich JK, Ley RE, Kao-Kniffin J. Selection on soil microbiomes reveals reproducible impacts on plant function. ISME J. 2015;9:980–9.
[101] Ludwig-Müller J. Plants and endophytes: equal partners in secondary metabolite production? Biotechnol Lett. 2015;37:1325–34.
[102] Hol WHG, de Boer W, Termorshuizen AJ, et al. Reduction of rare soil microbes modifies plant-herbivore interactions. Ecol Lett. 2010;13:292–301.
[103] Badri DV, Zolla G, Bakker MG, Manter DK, Vivanco JM. Potential impact of soil microbiomes on the leaf metabolome and on herbivore feeding behavior. New Phytol. 2013;198:264–73.
[104] Tomczak VV, Schweiger R, Müller C. Effects of arbuscular mycorrhiza on plant chemistry and the development and behavior of a generalist herbivore. J Chem Ecol. 2016;42:1247–58.
[105] van Elsas JD, Chiurazzi M, Mallon CA, Elhottova D, Kristufek V, Falcao Salles J. Microbial diversity determines the invasion of soil by a bacterial pathogen. Proc Natl Acad Sci. 2011;109:1159–64.
[106] Meiners SJ, Phipps KK, Pendergast TH, Canam T, Carson WP. Soil microbial communities alter leaf chemistry and influence allelopathic potential among coexisting plant species. Oecologia. 2017;183:1155–65.
[107] Busby PE, Soman C, Wagner MR, et al. Research priorities for harnessing plant microbiomes in sustainable agriculture. PLoS Biol. 2017;15:1–14.
[108] Mitter B, Pfaffenbichler N, Flavell R, et al. A new approach to modify plant microbiomes and traits by introducing beneficial bacteria at flowering into progeny seeds. Front Microbiol. 2017;8:1–10.
[109] Shelomi M, Lo W-S, Kimsey LS, Kuo C-H. Analysis of the gut microbiota of walking sticks (Phasmatodea). BMC Res Notes. 2013;6:1–19.
[110] Whitaker MRL, Salzman S, Sanders J, Kaltenpoth M, Pierce NE. Microbial communities of lycaenid butterflies do not correlate with larval diet. Front Microbiol. 2016;7:1–13.
[111] Tester M, Smith SE, Smith FA. The phenomenon of "nonmycorrhizal" plants. Can J Bot. 1987;65:419–31.
[112] Bunn R, Lekberg Y, Zabinski C. Arbuscular mycorrhizal fungi ameliorate temperature stress in thermophilic plants. Ecology. 2009;90:1378–88.
[113] Koziol L, Bever JD, Hawkes CV. Mycorrhizal response trades off with plant growth rate and increases with plant successional status. Ecology. 2015;96:1768–74.
[114] Gopal M, Gupta A. Microbiome selection could spur next-generation plant breeding strategies. Front Microbiol. 2016;7:1–10.
[115] Philippot L, Raaijmakers JM, Lemanceau P, van der Putten WH. Going back to the roots: the microbial ecology of the rhizosphere. Nat Rev Microbiol. 2013;11:789–99.
[116] Johnson NC, Graham JH, Smith FA. Functioning of mycorrhizal associations along the mutualism-parasitism continuum. New Phytol. 1997;135:575–86.
[117] Nouri E, Breuillin-Sessoms F, Feller U, Reinhardt D. Phosphorus and nitrogen regulate arbuscular mycorrhizal symbiosis in *Petunia hybrida*. PLoS One. 2014;9:e90841.
[118] Lehmann A, Barto EK, Powell JR, Rillig MC. Mycorrhizal responsiveness trends in annual crop plants and their wild relatives-a meta-analysis on studies from 1981 to 2010. Plant Soil. 2012;355:231–50.
[119] Kiers ET, Hutton MG, Denison RF. Human selection and the relaxation of legume defences against ineffective rhizobia. Proc Biol Sci. 2007;274:3119–26.

[120] Pérez-Jaramillo JE, Mendes R, Raaijmakers JM. Impact of plant domestication on rhizosphere microbiome assembly and functions. Plant Mol Biol. 2016;90:635–44.
[121] Lindenmayer D, Hobbs RJ, Montague-Drake R, et al. A checklist for ecological management of landscapes for conservation. Ecol Lett. 2008;11:78–91.
[122] Hakes AS, Cronin JT. Successional changes in plant resistance and tolerance to herbivory. Ecology. 2012;93:1059–70.
[123] Berg G, Smalla K. Plant species and soil type cooperatively shape the structure and function of microbial communities in the rhizosphere. FEMS Microbiol Ecol. 2009;68:1–13.
[124] Franklin RB, Mills AL. Multi-scale variation in spatial heterogeneity for microbial community structure in an eastern Virginia agricultural field. FEMS Microbiol Ecol. 2003;44:335–46.
[125] Cookson WR, Murphy DV, Roper MM. Characterizing the relationships between soil organic matter components and microbial function and composition along a tillage disturbance gradient. Soil Biol Biochem. 2008;40:763–77.
[126] Cai H, Ma W, Zhang X, et al. Effect of subsoil tillage depth on nutrient accumulation, root distribution, and grain yield in spring maize. Crop J. 2014;2:297–307.
[127] Ho A, Brink E van den, Reim A, Krause SMB, Bodelier PLE. Recurrence and frequency of disturbance have cumulative effect on methanotrophic activity, abundance, and community structure. Front Microbiol. 2016;6:1–10.
[128] Carson WP, Root RB. Herbivory and plant species coexistence: community regulation by an outbreak phytophagous. Ecol Monogr. 2000;70:73–99.
[129] Davidson DW. The effects of herbivory and granivory on terrestrial plant succession. Oikos. 2013;68:23–35.
[130] Lundberg DS, Lebeis SL, Paredes SH, et al. Defining the core *Arabidopsis thaliana* root microbiome. Nature. 2013;488:86–90.
[131] Edwards J, Johnson C, Santos-Medellín C, et al. Structure, variation, and assembly of the root-associated microbiomes of rice. Proc Natl Acad Sci. 2015;12:E911–20.
[132] Pendergast TH, Burke DJ, Carson WP. Belowground biotic complexity drives aboveground dynamics: A test of the soil community feedback model. New Phytol. 2013;197:1300–10.
[133] Gaiero JR, McCall CA, Thompson KA, Day NJ, Best AS, Dunfield KE. Inside the root microbiome: bacterial root endophytes and plant growth promotion. Am J Bot. 2013;100:1738–50.
[134] Hardoim PR, van Overbeek LS, Berg G, et al. The hidden world within plants: ecological and evolutionary considerations for defining functioning of microbial endophytes. Microbiol Mol Biol Rev. 2015;79:293–320.
[135] Moise ERD, Henry HAL. Like moths to a street lamp: exaggerated animal densities in plot-level global change field experiments. Oikos. 2010;119:791–5.
[136] Fry WE. *Phytophthora infestans*: new tools (and old ones) lead to new understanding and precision management. Annu Rev Phytopathol. 2016;54:529–47.
[137] Hoeksema JD, Chaudhary VB, Gehring CA, et al. A meta-analysis of context-dependency in plant response to inoculation with mycorrhizal fungi. Ecol Lett. 2010;13:394–407.
[138] Mishra Y, Johansson Jänkänpää H, Kiss AZ, Funk C, Schröder WP, Jansson S. Arabidopsis plants grown in the field and climate chambers significantly differ in leaf morphology and photosystem components. BMC Plant Biol. 2012;12:6.
[139] Agrawal AA, Hastings AP, Johnson MTJ, Maron JL, Salminen J. Insect herbivores drive real-time ecological and evolutionary change in plant populations. Science (80-). 2010;5317:113–6.
[140] Uesugi A, Kessler A. Herbivore exclusion drives the evolution of plant competitiveness via increased allelopathy. New Phytol. 2013;198:916–24.
[141] Kuzyakov Y, Domanski G. Carbon input by plants into the soil. Review. J Plant Nutr Soil Sci. 2000;163:421–31.

[142] Steinauer K, Chatzinotas A, Eisenhauer N. Root exudate cocktails: the link between plant diversity and soil microorganisms? Ecol Evol. 2016;6:7387–96.
[143] Jones DL, Jones DL, Hodge A, Kuzyakov Y. Plant and mycorrhizal regulation of rhizodeposition. New Phytol. 2004;163:459–80.
[144] Yergeau E, Sanschagrin S, Maynard C, St-Arnaud M, Greer CW. Microbial expression profiles in the rhizosphere of willows depend on soil contamination. ISME J. 2014;8:344–58.
[145] Ofek M, Voronov-Goldman M, Hadar Y, Minz D. Host signature effect on plant root-associated microbiomes revealed through analyses of resident vs. active communities. Environ Microbiol. 2014;16:2157–67.
[146] Peiffer JA, Spor A, Koren O, et al. Diversity and heritability of the maize rhizosphere microbiome under field conditions. Proc Natl Acad Sci. 2013;110:6548–53.
[147] Haney CH, Samuel BS, Bush J, Ausubel FM, Hospital MG. Associations with rhizosphere bacteria can confer an adaptive advantage to plants. Nat Plants. 2016;1:1–22.
[148] Wagner MR, Lundberg DS, del Rio TG, Tringe SG, Dangl JL, Mitchell-Olds T. Host genotype and age shape the leaf and root microbiomes of a wild perennial plant. Nat Commun. 2016;7:1–15.
[149] Truyens S, Weyens N, Cuypers A, Vangronsveld J. Bacterial seed endophytes: genera, vertical transmission and interaction with plants. Environ Microbiol Rep. 2015;7:40–50.
[150] Leff JW, Lynch RC, Kane NC, Fierer N. Plant domestication and the assembly of bacterial and fungal communities associated with strains of the common sunflower, *Helianthus annuus*. New Phytol. 2016;214:412–23.
[151] Schreiter S, Ding GC, Heuer H, et al. Effect of the soil type on the microbiome in the rhizosphere of field-grown lettuce. Front Microbiol. 2014;5:1–13.
[152] Chau JF, Bagtzoglou AC, Willig MR. The effect of soil texture on richness and diversity of bacterial communities. Environ Forensics. 2011;12:333–41.
[153] Berns AE, Philipp H, Narres HD, Burauel P, Vereecken H, Tappe W. Effect of gamma-sterilization and autoclaving on soil organic matter structure as studied by solid state NMR, UV and fluorescence spectroscopy. Eur J Soil Sci. 2008;59:540–50.
[154] Cipollini D, Rigsby CM, Barto EK. Microbes as targets and mediators of allelopathy in plants. J Chem Ecol. 2012;38:714–27.
[155] Uesugi A, Poelman EH, Kessler A. A test of genotypic variation in specificity of herbivore-induced responses in *Solidago altissima* L. (Asteraceae). Oecologia. 2013;173:1387–96.
[156] Yergeau E, Bell TH, Champagne J, et al. Transplanting soil microbiomes leads to lasting effects on willow growth, but not on the rhizosphere microbiome. Front Microbiol. 2015;6:1–14.
[157] Yan Y, Kuramae EE, Klinkhamer PGL, van Veen JA. Revisiting the dilution procedure used to manipulate microbial biodiversity in terrestrial systems. Appl Environ Microbiol. 2015;81:4246–52.
[158] Howard MM, Bell TH, Kao-Kniffin J. Soil microbiome transfer method affects microbiome composition, including dominant microorganisms, in a novel environment. FEMS Microbiol Lett. 2017;364:1–8.
[159] Wagner MR, Lundberg DS, Coleman-Derr D, Tringe SG, Dangl JL, Mitchell-Olds T. Natural soil microbes alter flowering phenology and the intensity of selection on flowering time in a wild *Arabidopsis* relative. Ecol Lett. 2014;17:717–26.
[160] Bell TH, Stefani FOP, Abram K, et al. A diverse soil microbiome degrades more crude oil than specialized bacterial assemblages obtained in culture. Appl Environ Microbiol. 2016;82:5530–41.
[161] Panke-Buisse K, Lee S, Kao-Kniffin J. Cultivated sub-populations of soil microbiomes retain early flowering plant trait. Microb Ecol. 2016;73:394–403.

[162] Folman LB, Postma J, Veen JA. Ecophysiological characterization of rhizosphere bacterial communities at different root locations and plant developmental stages of cucumber grown on rockwool. Microb Ecol. 2001;42:586–97.
[163] Tuckmantel T, Leuschner C, Preusser S, et al. Root exudation patterns in a beech forest: dependence on soil depth, root morphology, and environment. Soil Biol Biochem. 2017;107:188–97.
[164] Barillot CDC, Sarde CO, Bert V, Tarnaud E, Cochet N. A standardized method for the sampling of rhizosphere and rhizoplan soil bacteria associated to a herbaceous root system. Ann Microbiol. 2013;63:471–6.
[165] De Boer W, Hundscheid MPJ, Gunnewiek PJAK, et al. Antifungal rhizosphere bacteria can increase as response to the presence of saprotrophic fungi. PLoS One. 2015;10:1–15.
[166] Turner TR, Ramakrishnan K, Walshaw J, et al. Comparative metatranscriptomics reveals kingdom level changes in the rhizosphere microbiome of plants. ISME J. 2013;7:2248–58.
[167] Hagemeyer J. Ecophysiology of plant growth under heavy metal stress. In: Prasad MNV, editors. Heavy metal stress in plants. Berlin, Heidelberg, Germany: Springer; 2004. p. 201–22.
[168] Smith TA. Effects of disturbance on seed germination in some annual plants. Ecology. 1970;51:1106–8.
[169] Meyer AH, Schmid B. Experimental demography of rhizome populations of establishing clones of *Solidago altissima*. J Ecol. 1999;87:42–54.
[170] Takeno K. Stress-induced flowering: the third category of flowering response. J Exp Bot. 2016;67:4925–34.
[171] Potts SG, Biesmeijer JC, Kremen C, Neumann P, Schweiger O, Kunin WE. Global pollinator declines: trends, impacts and drivers. Trends Ecol Evol. 2010;25:345–53.
[172] Dudareva N, Negre F, Nagegowda DA, Orlova I. Plant volatiles: recent advances and future perspectives. 2006;25:417–40.
[173] Aartsma Y, Bianchi FJJA, van der Werf W, Poelman EH, Dicke M. Herbivore-induced plant volatiles and tritrophic interactions across spatial scales. New Phytol. 2017;216:1054–63.
[174] Ugrekhelidze D, Korte F, Kvesitadze G. Uptake and transformation of benzene and toluene by plant leaves. Ecotoxicol Environ Saf. 1997;37:24–9.
[175] Tangahu BV, Rozaimah S, Abdullah S, et al. A review on heavy metals (As, Pb, and Hg) uptake by plants through phytoremediation. Int J Chem Eng. 2011;2011:1–31.
[176] Tenenboim H, Brotman Y. Omic relief for the biotically stressed: metabolomics of plant biotic interactions. Trends Plant Sci. 2016;21:781–91.
[177] Tholl D, Boland W, Hansel A, Loreto F, Röse USR, Schnitzler JP. Practical approaches to plant volatile analysis. Plant J. 2006;45:540–60.
[178] Morrell K, Kessler A. Plant communication in a widespread goldenrod: keeping herbivores on the move. Funct Ecol. 2016;31:1049–61.
[179] Kessler A, Baldwin IT. Defensive function of herbivore-induced plant volatile emissions in nature. Sci. 2001;291:2142–4.
[180] Li T, Blande JD, Holopainen JK. Atmospheric transformation of plant volatiles disrupts host plant finding. Sci Rep. 2016;6:33851.
[181] Blande JD, Holopainen JK, Niinemets Ü. Plant volatiles in polluted atmospheres: Stress responses and signal degradation. Plant Cell Environ. 2014;37:1892–904.
[182] Hobbs PJ, Webb J, Mottram TT, Grant B, Misselbrook TM. Emissions of volatile organic compounds originating from UK livestock agriculture. J Sci Food Agric. 2004;84:1414–20.
[183] Wilson JK, Kessler A, Woods HA. Noisy communication via airborne infochemicals. Bioscience. 2015;65:667–77.
[184] Parker DB, Malone GW, Walter WD. Vegetative environmental buffers and exhaust fan deflectors for reducing downwind odor and VOCs from tunnel-ventilated swine barns. Trans ASABE. 2012;55:227–40.

[185] De Vos M, Jander G. Choice and no-choice assays for testing the resistance of *A. thaliana* to chewing insects. J Vis Exp. 2008. DOI: 10.3791/683.

[186] Bode RF, Kessler A. Herbivore pressure on goldenrod (*Solidago altissima* L., Asteraceae): its effects on herbivore resistance and vegetative reproduction. J Ecol. 2012;100:795–801.

[187] Li Y-P, Zhang J, Chen Y, et al. Changes in soil microbial communities due to biological invasions can reduce allelopathic effects. J Appl Ecol. 2017;54:1281–90.

[188] Vazquez-de-Aldana BR, Romo M, Garcia-Ciudad A, Petisco C, Garcia-Criado B. Infection with the fungal endophyte Epichloë the allelopathic potential of red fescue. Ann Appl Biol. 2002;159:281–90.

Don Cowan, Evelien Adriaenssens, Pieter De Maayer, Pedro Lebre, Thulani Makhalanyane, Jean-Baptiste Ramond, Marla Trindade, Angel Valverde and Surendra Vikram

5 Metagenomics of extreme environments: methods and applications

5.1 Introduction

Metagenomics, with all its associated technologies, has completely revolutionized microbial ecology in little more than a decade and a half. This revolution has applied to extreme environments as much as any other, and possibly more so due to the scientific community's enduring fascination with these unusual habitats. From the earliest phylogenetic surveys of single sites and samples, laboriously constructed using, at best, a few hundred dereplicated 16S rRNA gene library clones, and where inserts were slowly sequenced using ABI platforms, to the current status where a "standard" phylogenetic survey may involve hundreds of samples and multiple replicates sequenced on a next generation sequencing (NGS) platform over a few hours or days, we can easily identify the critical drivers of the "revolution." The principal drivers are technological and economic, with a combination of new sequencing platforms capable of multiple orders-of-magnitude greater capacity and speed, and the inversely correlated reduction in sequence costs (as per the now famous nucleic acid sequencing cost time-line graph: the Moore's Law of Metagenomics [www.genome.gov]). The other major driver has been the spectacular development of new software packages (and of their user-friendliness) in direct response to the needs of the research community, which has placed the capacity to undertake high-sequence-volume and sophisticated metagenomics projects within the grasp of any reasonably supported laboratory anywhere in the world. It is no coincidence that most of the references to the software packages identified in this chapter have publication dates since 2012!

We have seen a series of landmark studies of extremophile metagenomics over this relatively short period: the (oligotrophic) Sargasso Sea metagenome [1], the genome reconstruction and physiological interpretation of the acid mine drainage metagenome [2], the genome-based re-assembly of the "tree of life" (in which the extremophilic taxa feature as dominant clades), and the large-scale reconstruction of microbial genomes as a tool for understanding community function [3]. These studies (which represent just single points in a developing research continuum) are clear evidence of the dramatic rise and expansion of metagenomics technologies, to a point where it might be concluded that classical culture-dependent microbiology is a "thing of the past." We argue strongly, as have many others, that this is not the case: modern metagenomics is, at least in part, predictive (as in the interpretation

https://doi.org/10.1515/9783110525786-005

of genome content on the context of metabolic function). At the current state of the technology, such predictions still require experimental verification, parts of which require the more classical methods of experimental microbiology (see Chapter 1).

5.2 Metagenomic DNA extraction

For metagenomics studies involving NGS technologies, the critical first step is to obtain high-quality metagenomic DNA (mDNA), both in terms of purity and concentration. When working with extreme environments, this can easily become a challenge. These systems generally contain limited microbial biomass when compared to more mesic systems, and "extreme mDNA" is often coextracted with PCR-inhibitory substances, such as metals (e.g. hydrothermal vents or mines) and/or salts (e.g. hypersaline lakes and halites), which can hamper downstream molecular analyses.

Two strategies can be employed to extract environmental mDNA. *Direct extraction*, which is the most commonly used in metagenomics studies, involves the lysis of (microbial) cells within their environmental matrix. For *indirect extraction*, cells are selectively removed from the matrix prior to their lysis [4]. For metagenomics, the issue with the direct extraction, particularly when studying extreme environments, is the coextraction of PCR inhibitors, which then requires additional purification steps and lower mDNA yield. Conversely, indirect extraction is time-consuming, requires specific equipment, and can be biased toward the isolation/selection of particular microbial cells [4, 5]. Nevertheless, both have been applied successfully to extreme environments [e.g. 5, 6]. In this section, we summarize the successful *tricks* developed to improve mDNA recovery from extreme environmental samples.

5.2.1 Direct mDNA extractions in extreme environments

Direct mDNA extractions can be performed using commercial kits [7] and the more classical CTAB- or phenol-based extraction procedures [e.g. 8, 9]. Both are still being evaluated by the Extreme Microbiome Project (launched in 2014; http://extrememicrobiome.org/), which further highlights the uniqueness of extreme environments when it comes to accessing their indigenous microbial genetic material.

To obtain directly extracted high-quality *extreme mDNA*, two strategies can be employed, and examples of their effective applications are given in Tab. 5.1: either the environmental samples are treated before the extraction (Tab. 5.1, pretreatment section) or the coextracted contaminants are removed *a posteriori* from *extreme DNA* mixtures by adding extra purification steps (Tab. 5.1, posttreatment section). Pretreatments are typically adapted to the origin of the samples, while posttreatment strategies all aim to remove coextracted contaminants (Tab. 5.1).

Tab. 5.1: Succesfull pre- and post-treatments for metagenomic DNA extraction of extreme environments

Treatment	Extreme environment	Environmental characteristics	Strategy	Important steps	References
Pre-treatment	Acid mine Drainage	Low pH (~0,7–1,3) and Iron-rich	Wash steps to exclude iron and raise pH	Wash buffer 1: PBS (pH of the sample of origin) Wash buffer 2: one part 2X buffer A (200 mM Tris [pH 8.0], 50 mM EDTA, 200 mM NaCl, 2 mM sodium citrate, 10 mM $CaCl_2$) and one part 50% glycerol Resuspension buffer: 2 X Buffer A	Bond et al, 2000 [8]
	Alkaline soils	pH 7,6–11 and humic acid-rich	Humic acid free high molecular weight mDNA	Activated charcoal (0.4 g), Proteinase K (0.2 mg), and 2 mL extraction buffer (N,N,N,N-cetyltrimethylammonium bromide 1%, polyvinylpolypyrrolidone (PVPP) 2%, 1.5 M NaCl, 100 mM EDTA, 0.1 M TE buffer (pH 8.0), 0.1 M sodium phosphate buffer (pH 8.0), and 100 μ L RNase A)	Verma and Satyanarayana, 2011 [179]
	Volcanic Environments	Clay-rich Andisol	Addition of DNA adsorption competitors	Skimmed milk or RNA treatment prior to extraction with a commercial kit	Herrera and Cockell, 2007 [5]
	Industrial waste (residue and lixiviate from chromite processing industry)	Hyper-alkaline (pH 9–14), hyper-saline (~100PSU) and metal rich (5–10 g of Chromium per kg soil and 2 to 108 g Iron per kg soil)	Remove cations and metals which inhibit Taq polymerase	3 washes with phosphate buffer or neutralization (wash with phosphate buffer until neutral pH is reached) in addition with sample crushing in liquid nitrogen	Brito et al, 2013 [180]
	Hot desert soils	High Salt and low biomass soils	Remove PCR inhibitors and extract sufficient mDNA for downstream applications	3 washes with TE buffer (10 mM Tris-EDTA, pH 5.0)	Scola et al, accepted for publication [181]

Tab. 5.1 (continued)

Treatment	Extreme environment	Environmental characteristics	Strategy	Important steps	References
	Contaminated river sediments	Complex organic pollutants and heavy metal contaminated	Remove PCR inhibitors	Multiple washes (until supernatant becomes clear) with a solution containing 0.1 $mol.L^{-1}$ EDTA, 0.1 $mol.L^{-1}$ Tris (pH 8.0), 1.5 $mol.L^{-1}$ NaCl, 0.1 $mol.L^{-1}$ NaH_2PO_4 and Na_2HPO_4	Fang et al, 2015 [182]
	Yellowstone National Parl thermal fixtures	pH 2.49–9.08/Temperature 42.4–90.6°C	Increase retrieved microbial diversity	Prior to DNA extraction, sample preservation in Sucrose Lysis Buffer [SLB]: 20 mM EDTA, 200 mM NaCl, 0.75 M sucrose, 50 mM Tris-HCl, pH 9.0	Mitchell and Takacs-Vasbach, 2008 [183]
	Deap subseafloor	Low biomass environment	Increase mDNA quality [size] and yield and microbial community diversity	Hot (70°C) alkaline lysis solution (1 M NaOH, 5 mM EDTA (pH 8.0), and 1% SDS) pretreatement	Morono et al, 2014 [184]
	Atmosphere	Extremely low biomass	Extract sufficient mDNA for high throughput sequencing	1. Sample collection on Tissuquartz filters 2. Treat a quarter of the filter with 1X PBS buffer 3. Centrifugation 3h 200 g 4°C 4. Concentrate microbial biomass by filtration on 0.2 μm filter.	Jiang et al, 2015 [185]
Post-treatment	Organic rich sediments	Silty clay-sediments with high organic content	Remove co-extracted contaminants to obtain PCR-amplifiable mDNA	Hydroxyapatite spin columns	Purdy et al, 1996 [186]
	Hot springs sediments from Zambia, China, New Zealand and Kenya	pH 5.0–10.0/Temperature 39.4–86.5°C/Cl concentrations of to 50875 mg Cl-/kg		PVPP (polyvinylpolypyrrolidone) spin columns	Valverde et al, 2012 [187]
	Low Biomass carbonate rock	Remove calcium inhibitors of PCR amplification		Dialysis	Barton et al, 2006 [188]
	Brackish contaminated sediments	Heavy metal contaminated estuarine sediments		Elutip-D column	Ramond et al, 2008 [189]
	Subsurface (45 m) marine sediments	Organic rich sediments		Cesium Chloride(CsCl) density gradient centrifugation	Juniper et al, 2001 [190]

5.2.2 Indirect mDNA extractions

Indirect mDNA extractions seem the most appropriate methods for metagenomic studies of extreme environments as only intact microbial cells are isolated and concentrated prior to their lysis, which ultimately reduces the coextraction of PCR inhibitors (and extracellular DNA). Their major drawback, however, remains their low mDNA recovery yields, even when working in non-extreme systems [27]. Nevertheless, these protocols enable the extraction of high-quality and high-molecular-weight *extreme DNA*, which makes it an ideal approach for functional metagenomics studies [27].

Indirect mDNA extraction procedures include the blending, the cation-exchange, and the Nycodenz® density gradient methods [6, 27, 28]. The blending protocol involves the mechanical release of microbial cells from their environmental matrix (e.g. soil particles) in a blending buffer (100 mM Tris-HCl, 100 mM sodium EDTA, 0.1% SDS, 1% CTAB, pH 8) and use of a blender, after which the cells are harvested by centrifugation [27]. The cation-exchange resin protocol combines slow centrifugation to remove coarse particles, and a chelating resin (e.g. Chelex 100) is added to eliminate contaminating and PCR-inhibiting ions (e.g. metals) [29]. Bacterial cells in the Nycodenz® gradient centrifugation approach are also separated from their environment by multiple centrifugation and washing steps [6].

Finally, single-cell genomics methods, i.e. where microbial genomes are sequenced one cell at a time (after isolation with fluorescence activated cell sorting, for example), fall within the indirect mDNA extraction approaches. This is a highly promising methodology for accessing novel genomes [30].

5.3 Microbial diversity analysis: phylogenetics

One of the principal ways by which microbial communities can be characterized is by quantifying their phylogenetic diversity. Currently, amplicon sequencing, usually targeting the small-subunit ribosomal RNA (16S) locus, is the method of choice for phylogenetic surveys, although shotgun metagenomics can also be used.

In general, a standard microbial diversity analysis can be divided into two parts: “upstream” and “downstream” analysis. The upstream analysis includes the processing of the raw data (sequencing output) and the generation of the key files (operational taxonomic unit [OTU] table and phylogenetic tree) for microbial analysis. The downstream analysis uses the OTU table and phylogenetic tree generated in the upstream step to perform diversity analysis, statistics, and visualizations of the data. In this section, we briefly describe these analyses and provide information on the software that can be used to perform them.

5.3.1 Upstream analysis

The upstream analysis workflow starts with the sequencing output (i.e. fastq files) and a user-generated mapping file, which contains information about each sample. The main information in this file is a unique identifier for each sample, the barcode used for each sample, the primer sequence used, and a description for each sample. Additional information necessary to interpret the results, such as which habitat the sample was taken from and environmental variables relevant to the study, can also be incorporated.

Demultiplexing and quality filtering

High-throughput sequencing allows multiple samples to be combined in a single sequencing run. However, each sequence must then be linked back to the individual sample that it came from via a DNA barcode. The barcodes, which are short-DNA sequences unique to each sample, are incorporated into each sequence from a given sample during PCR. Pipelines use the barcodes in the mapping file to demultiplex; that is, to assign the sequences back to the samples they belong to. Illumina NG sequencing technology generates *fastq* files that are already separated by sample. Most sequencing platforms generate a quality score for each nucleotide, associated to the probability that each nucleotide was read incorrectly. Pipelines use this quality score and user-defined parameters [31] to remove sequence reads that do not meet the desired quality.

OTU generation

The next step is to cluster the preprocessed sequences into OTUs, which is done based on sequence identity. In other words, sequences are clustered together if they are more similar than a user-defined identity threshold, presented as a percentage. This level of threshold is traditionally set at 97% of sequence similarity, conventionally assumed to represent bacterial species [32]. There are different approaches for OTU picking (e.g. *de novo*, closed reference, and open reference) and multiple algorithms for each of these approaches. The *de novo* approach groups sequences based on sequence identity [33]. The closed-reference approach matches sequences to an existing database of reference sequences [34]. If a sequence fails to match the database, it is discarded. The open-reference approach also starts with an existing database and tries to match the sequences against it. However, if a sequence does not match the database, it is added to the database as a new reference sequence [34].

Of the various algorithms available, the furthest-neighbor, average-neighbor, or nearest-neighbor methods are the most widely used [35]. Because these three algorithms are based on hierarchical clustering, they require loading of the distance matrix into computer memory and are therefore challenging to apply to large datasets. A solution to the distance matrix problem comes from *usearch* [36], or its open-source alternative *vsearch* [37], which are greedy algorithms based on using a single centroid in each OTU. These algorithms are much more efficient than hierarchical methods and they do not require a large distance matrix to be loaded into memory. Once the sequences have been clustered into OTUs, a representative sequence is picked for each OTU. The entire cluster will thus be represented by a single sequence, speeding up subsequent steps. Most pipelines allow the representative sequence to be selected using different techniques, such as choosing a sequence at random, choosing the longest sequence, or choosing the most abundant sequence or the first sequence.

Identify chimeric sequences

During the PCR amplification process, some of the amplified sequences can be produced from multiple parent sequences, generating sequences known as chimeras. The removal of chimeric sequences is important because they can artificially inflate diversity estimates. Two of the most common methods for detecting chimeras are *ChimeraSlayer* [38], which uses BLAST to identify potential chimera parents, and *uchime* [39], which can perform *de novo* chimera detection based on abundances as well as reference-based chimera detection.

Taxonomic assignment

The next step is to assign the taxonomy to each representative sequence. When using a closed-reference approach for OTU picking, the taxonomy of the sequences can be pulled out from the reference set. In case of the open-reference and *de novo* approaches, because the clusters are not created from any reference database, the taxonomy should be assigned using a separate reference database such as *RDP* [40], *Silva* [41], and *GreenGenes* [42].

Sequence alignment

Sequences must be aligned to infer a phylogenetic tree, which is used for diversity analyses and to understand the relationships among the sequences in the sample. Current methods for performing sequence alignment include, among others, *PyNAST* [43], *clustalw* [44], and *muscle* [45].

Phylogenetic construction

This step infers a phylogenetic tree from the multiple sequence alignment generated by the previous step. The phylogenetic tree represents the relationships among sequences in terms of the amount of sequence evolution from a common ancestor. This phylogenetic tree is used in many downstream analyses, such as the UniFrac metric for beta-diversity. Some current methods for inferring a phylogenetic tree are *FastTree* [46], *raxml* [47], and *muscle* [45].

Downstream analysis

Once the OTU table has been obtained, the recommendation is to perform a second quality-filtering step. For example, OTUs with a relative abundance lower than 0.005% of the total number of sequences can be discarded [31]. This step greatly reduces the problem of false OTUs, most of which are present at very low abundance. A further step is to equalize (rarefy) the sampling depth; that is, equalize the number of sequences that should be included in each sample for diversity analyses. Rarefaction has been traditionally recommended because many of the commonly used diversity metrics are very sensitive to the number of sequences obtained in a given sample. The optimal sampling depth is data dependent. For instance, if most samples have more than 20,000 sequences and the remainder range from 2,000 to 5,000 sequences per sample, it would be recommended to use 10,000 as the rarefaction level. Nevertheless, more recently, it has been suggested that rarefaction should be avoided, especially if the study attempts to detect differential abundance of OTUs between predefined groups of samples [48].

Taxon summaries

One way to visualize the OTUs in each sample is to summarize the relative abundance of the taxa present in a set of samples on multiple taxonomic levels (e.g. phylum, order, etc.). This provides a quick way to identify samples that may be drastically different from others (i.e. outliers) and to visually identify expected patterns and differences between and among samples. For example, this tool was used to identify how soil microbial communities respond to the presence of an ancient relic in the Antarctic Dry Valleys [49].

5.3.2 Diversity analysis

Microbial diversity can be measured in many ways, but three main distinctions are common: alpha-, beta-, and gamma-diversities. Alpha-diversity is defined as the diversity of organisms in a given sample or environment. Beta-diversity is the difference in diversities across samples or environments. Gamma-diversity measures the diversity at a broader scale, such as a province or region.

Alpha-diversity analysis

Most studies of microbial community diversity to date have used species-based measures of diversity. Both qualitative species-based measures, such as Chao 1 or ACE, and quantitative species-based measures, such as the Shannon or Simpson indices, have been widely applied [50]. Nevertheless, because microbial community diversity is usually assessed using molecular data, one can account for the phylogenetic relationships among species. One of the most used metrics incorporating this phylogenetic information is the phylogenetic distance (PD) [51]. It is important to note that the choice of metric will depend on the question to be interrogated. For instance, one might be interested in pure estimates of community richness (such as the observed number of OTUs), in pure estimates of evenness, or of measures that combine richness and evenness such as Shannon entropy. Measuring alpha diversity is important for comparing the total diversity in different communities, for example, to show that plant-free cold desert soils typically have the lowest levels of phylogenetic and taxonomic diversity in comparison with other biomes [52] and that at temperatures between 7.5 and 99°C, species richness and diversity indices peak at 24°C and decrease at higher and lower temperature extremes [53].

Beta-diversity analysis

As with alpha-diversity, there are many phylogenetic and nonphylogenetic beta-diversity metrics. However, the two most commonly used classes of measures of beta diversity are (i) the classical metrics, calculated directly from measures of gamma and alpha diversity, and (ii) multivariate measures, based on pairwise resemblances (similarity, dissimilarity or distance) among sample units [54]. Among the first group is the so-called Whittaker beta-diversity, which is the number of times by which the richness in a region is greater than the average richness in the smaller-scale units. In contrast, the latter group is related to the well-known Jaccard, Bray-Curtis, and UniFrac (Unique Fraction metric) coefficients. Unifrac measures the PD between sets of taxa in a phylogenetic tree as the fraction of branch length of the tree that leads to descendants from either one community or the other but not both [55]. Unweighted Unifrac only considers the absence/presence of the OTUs, whereas the weighted Unifrac metric is weighted by the difference in probability mass of OTUs from each community for each branch. Weighted Unifrac is thus recommended for detecting community differences that arise from differences in relative abundance of taxa, rather than in which taxa are present. Like other metrics considering taxon abundance, weighted Unifrac is sensitive to the bias from DNA extraction efficiency, PCR amplification, etc. The choice of metrics is critical in beta-diversity analysis as metrics differ substantially in their ability to detect clustering or gradient patterns among microbial communities on the same dataset [56]. Beta diversity analysis has revealed, for example, that permafrost microbial communities respond rapidly to thaw [57] and that pH is primarily responsible for structuring whole microbial (bacterial and archaeal) communities in extreme and heterogeneous mine tailings [58].

Statistical significance of differences in alpha- and beta-diversity

Knowing which statistical tests should be applied depends on the particular hypotheses and predictions defined *a priori* in a particular study. For alpha-diversity, probably the most common test is the analysis of variance (ANOVA), which is an extension of the familiar *t*-test that is used to compare the means of two groups. Comparisons between distance matrices can be performed using the Mantel test and the partial Mantel test [59]. The Mantel test is a nonparametric test that compares two distance matrices and calculates a correlation coefficient and a significant *p*-value using permutations. The partial Mantel test is similar to the Mantel test, except that the analysis is controlled by a third variable. Other multivariate analyses for exploring significant relationships between the beta-diversity distance matrix and factors or covariates are ANOSIM [60] and PERMANOVA [61]. ANOSIM is a nonparametric statistical test that compares ranked beta-diversity distances between different groups. PERMANOVA partitions the variance in a similar way to the ANOVA family of tests, specifically testing if the variation within a category is smaller or greater than variation between categories.

OTU category significance

It is often appropriate to compare two or more groups for differences in the abundance of OTUs. One way to make such as assessment is to compare the relative abundances of each microbial member between the two groups. Classical methodologies include the use of ANOVA and nonparametric Kruskal-Wallis tests, but indicator species analysis [62] and linear discriminant analysis (LDA) effect size (LEfSe) [63] may also be used.

While many of these taxa may be significantly different between groups according to the raw *p*-value, it is extremely important that only *p*-values that have been corrected-for against multiple comparisons are considered as significant. Typically, an OTU table contains hundreds or thousands of OTUs, and thus a *p*-value is likely to reach significance based solely on the large number of statistical comparisons being computed. This type of analysis has shown, for instance, that cyanobacteria are among the most important functional groups in hot deserts hypolithic communities [64].

5.3.3 Software

There are different bioinformatic and statistical tools to efficiently and reproducibly perform the analyses described above. For upstream analyses, these include, among others, *QIIME* [65], *mothur* [35], *MG-RAST* [10], and the *RDP* pipeline [40]. Downstream analysis can be implemented, for example, with *R* packages such as *Vegan* [66], *ade4* [67], *ape* [68] and *phyloseq* [69], but also with more user-friendly software packages such as *PAST* (https://folk.uio.no/ohammer/past/) and *PRIMER-E* (http://www.primer-e.com/index.htm).

5.4 Metaviromics

Determining the metagenome content of viral communities in extreme environments poses additional challenges to those described earlier in this chapter. Viral genomes are generally much smaller than those of cellular organisms, requiring a viral isolation or enrichment step before nucleic acid extraction. In addition, many extreme environments have a low microbial biomass, further complicating extraction procedures for metaviromic sequencing. While it is possible to get viral signatures for specific groups or families from amplicons using group-specific marker gene primers (reviewed in Ref. [70]), there is no such thing as a universal viral marker gene [71], which makes (meta)viromics the only way to study the viral community as a whole.

5.4.1 Viral-like particle extraction or enrichment

The first step in viromics is generally the extraction or enrichment of viral-like particles (VLPs) in the sample material. The purpose of this step is to remove cellular organisms and other potential contaminants and bring the volume of the sample down to a manageable volume for the next step. For aqueous samples coming from environments such as hot springs or (hyper)saline lakes or salterns, the most commonly used technique is tangential flow filtration (TFF) [72]. Using a filtration cartridge with a molecular weight cutoff of 30 or 100 kDa, large sample volumes (generally more than 10 l) can be reduced while retaining the viral fraction which is larger than the pore size [73–79]. Alternatively, for smaller volumes, spin filter columns with similar pore sizes that fit in a normal benchtop centrifuge can be used for volume reduction [80]. The volume reduction step can either be preceded by a filtration or centrifugation step to remove cellular particles [73, 75–78] or followed by it [74, 79]. Removal of the cellular fraction before TFF reduces the potential clogging of the filter cartridge but requires larger volumes to be pretreated. An alternative to TFF is the use of chemical flocculation with $FeCl_3$, which precipitates VLPs from aqueous solutions onto a filter membrane [81, 82].

Soil (e.g. hyperarid desert soil) or sediment (e.g. hadopelagic sediment from deep see trenches) samples are generally mixed with sterile water or buffer to suspend the VLPs and to dislodge them from the soil or substrate matrix. Different approaches have been used to generate soil suspensions, such as mixing with a potassium citrate buffer followed by centrifugation and filtration [83–85], suspension in large volumes of water followed by TFF [86–88], or suspension in high-saline solutions followed by filtration [89, 90].

Further volume reduction and enrichment of VLPs can be achieved by PEG 8000 (polyethylene glycol of molecular weight 8000) precipitation, spin filtration, and/or ultracentrifugation at >15,000 × g. Additional purification of the samples can be done with CsCl density gradient centrifugation [89–92]. For samples with high salt concentrations, a buffer exchange step can be introduced [74].

5.4.2 Nuclease treatment

Including a nuclease treatment (DNase and/or RNase) before nucleic acid extraction reduces contamination with cellular material [93]. This requires a viral suspension that is relatively pure (i.e. devoid of enzyme inhibitors) and has an Na^+ ion concentration that is sufficiently low so as not to inhibit nuclease activity. For "dirtier" samples, benzonase has been used to degrade contamination nucleic acid [79, 94], but this can potentially reduce the abundance of certain viral particles [95]. Another approach has been to mix the viral suspensions with low melting agarose and perform the DNase treatment in agarose plugs [77, 78].

5.4.3 Nucleic acid extraction and amplification

RNA, DNA, or a combination of both can be extracted from the VLP suspensions using commercial kits or more traditional phenol:chloroform-based extraction methods. DNA extraction using the latter has been described in detail by Vega Thurber and colleagues [72]. In general, the viral suspension is mixed into a buffer containing EDTA and a proteinase. The nucleic acid is then extracted using a mixture of phenol:chloroform (1:1) or acid phenol for phase separation. This is usually followed by ethanol or isopropanol precipitation. When RNA is targeted, an additional DNase treatment (with RNase-free DNase) can be performed to remove contaminating viral DNA [80].

In many cases, the yield of the nucleic acid is too low for direct sequencing library preparation. The most commonly used approach in viromics is currently *multiple displacement amplification* (MDA) with the phi29 polymerase. This method, however, has been shown to be biased for small, circular ssDNA viruses and should therefore be used with caution [96–98]. Another amplification method can be used as an alternative: random-priming sequence independent single primer amplification (RP-SISPA). This method is compatible with cDNA synthesis steps of RNA viromes but can be biased toward the most abundant members of the community [99–102].

5.5 De novo metagenome assembly (the genome-centric approach)

Metagenomic data can be analyzed by two distinct approaches, namely "gene-centric" or "genome-centric" approaches [103]. The gene-centric method of metagenomic analysis involves the homology-based assignment of functions to reads or open reading frames obtained from metagenomics assemblies, although these short reads are of limited value for taxonomic assignments [103]. One of the major challenges in microbial ecology is to study the uncultured microorganisms that make up a major part of the global fraction of organisms in any sampled environment and almost certainly have important roles in these environments. The genome-centric metagenomics approach provides a plausible and promising means for the identification and characterization of the genomes of uncultured microorganisms.

The first integral step of genome-centric metagenomics is the generation of accurate assemblies of the sequence data. Over the past few years, advanced NGS technologies have been developed that are variously capable of producing very deep coverage of genomic data, long read lengths, and/or low sequencing error rates. The data obtained from the sequencing platform must first be analyzed and processed to attain the high-quality short reads prior to assembly. Various tools for quality analysis are available, such as *FastQC*, *Adapterremoval*, *Prinseq*, the *NGS QC toolkit*, and *trimmomatic*; each has characteristic features for the quality processing of data [104–108]. A Phred quality score, first developed for base-calling during the Human Genome Project, is a measure of the quality of base-calling from the sequencing platform [109].

The assembly of metagenomes is far more complex and challenging than genome assembly, since most environmental samples contain varying abundances of closely related species, which results in differential coverage/depth of sequencing data [110]. Due to the increasing demand for appropriate assembly tools, a number of assembly algorithms have been designed specifically for metagenome assemblies; these include *metaSPAdes* [111], *MEGAHIT* [112], *IDBA-UD* [113], and *RayMeta* [114].

The selection of a metagenomic assembly algorithm is highly dependent on the metagenomics data to be analyzed [115]. In particular, the complexity of the microbial community structure is a critical factor in determining the assembly performance of any software [116]. There is, however, no universal rule for the detection of optimal (or erroneous) assemblies. A software tool, *MetaQUAST*, has been developed that can compare assemblies from different assembly platforms and can provide feedback on the quality of the assembly generated [117].

The process of complete genome assembly from metagenomic sequence datasets involves the clustering of contigs into different taxonomic bins, where each bin effectively represents a draft genome [118]. Several genomic and sequence data signatures have been used to bin metagenomic contigs into their distinct genomes [118]. Frequently used signatures include tetranucleotide or oligonucleotide frequencies [119–121], G+C content [122, 123] and sequence coverage [124, 125]. Alternatively, it is possible to combine these signatures in a human-guided manual mode [126, 127]. These approaches require high levels of bioinformatics expertise and are time-consuming. In consequence, recent emphasis has been placed on the development of efficient binning algorithms, such as *CONCOCT* [128], *GroopM* [129], *Metabat* [130], *Maxbin2* [131], and *DAS tools* [132], which automate and accelerate genome reconstruction from metagenomics contigs.

The genome bins should be assessed for completeness and contamination based on the presence of known universal single copy genes (SCGs) for archaea and bacteria [126]. An automated post-binning pipeline, *Checkm*, for the quality assessment of the reconstructed bins, analyzes and calculates completeness, contamination, and strain heterogeneity [133]. Other studies have identified SCGs for both bacteria and archaea, to use as benchmarks in the quality assessment of genome reconstruction [134, 135].

Genome reconstruction strategies have been applied to metagenome sequence datasets obtained from a wide range of environments, including extreme habitats such as hypersaline lakes [136], cold environments [137, 138], and deep sea hydrothermal vents [139, 140], and have enabled researchers to identify and explore novel bacterial and archaeal lineages within these environments. For example, metagenomic binning of sequences from a hypersaline lake revealed a novel class of Euryarchaeota, the "Nanohaloarchaea" [136]. Similarly, a metagenomic-based genome reconstruction depth profile sample from the Mediterranean Sea produced a number of novel marine group euryarchaeota (MG-III) from both the photic and deep-sea zones, and these genomes were found to harbor genes responsible for photoheterotrophic metabolism [140]. Recently, an extensive study reported the identification of 117 new bacterial phylum-level lineages from the sediment and ground water metagenomes collected adjacent to the Colorado River (USA) [3].

In summary, genome-centric approaches are increasingly popular for the discovery of uncultured bacteria and archaea. This approach has no substitute in cases where the recovery of culturable microorganisms is not possible, but where DNA can be easily captured for shotgun sequencing [141]. Furthermore, unlike 16S rRNA and targeted amplicon sequencing, genome-centric metagenomics represents a means of exploring both phylogenetic diversity and the functional capacity of an environmental microbiota [142].

5.6 Exploring the functional ecology of metagenomes

5.6.1 The use of bioinformatics pipelines for predicting functional capacity

One of the major objectives of microbial ecology is to determine the functional capacity of a given community and how this capacity varies with taxonomic composition and abiotic factors present in the environment. This functional analysis relies primarily on powerful computational tools that can process large datasets and output information that can be used to infer functional capacity and variation for any given metagenomic sample [143]. Currently, there are a number of open source bioinformatics pipelines (Tab. 5.2) for the functional annotation and prediction

Tab. 5.2: Some of the principal open-source bioinformatic pipelines routinely used to predict functionality from metagenomics sequence data, together with the protein and gene databases accessed by each pipeline.

Pipelines	Functional annotation databases	URL
MG-RAST [10]	SEED framework [11]	http://metagenomics.anl.gov/
IMG/M [12]	NCBI NR/Refseq [13] KEGG [14] COGS [15] InterPro [16] IMG pathways [17]	https://img.jgi.doe.gov/mer/
EBI metagenomics [18]	Gene 3D [19] PRINTS [20] Pfam [21] TIGRFAMs [22] PROSITE [23] Gene Ontology (GO) [24]	https://www.ebi.ac.uk/metagenomics
WebMGA [25]	COGS KOGs Pfam TIGRFAMs Gene Ontology (GO)	http://weizhongli-lab.org/ metagenomic-analysis
MEGAN [26]	NCBI NR	http://www.ab-informatik. uni-tuebingen.de/software/megan

of metagenomic sequence datasets, all of which are supported by an increasing number of online databases of reference proteins (e.g. *NCBI Refseq* and *NR*), protein families (e.g. *Pfam*), orthologous groups (*COGs, NOGs*), metabolic and cellular pathways (e.g. *KEGG*), as well as plethora of online resources for prediction of cellular localization, specific enzymatic activity and virulence [144]. *MG-RAST*, one of the most widely using functional annotation pipelines for genomic and metagenomics data, uses a homology-based approach in which query sequences are blasted against the SEED nonredundant protein database and the *FIGfams* protein family collection and are consequently clustered into functional families following the SEED subsystem annotation framework [10]. In turn, the SEED framework is a platform-independent, manually curated database in which proteins are classified and clustered according to their functional roles into a hierarchy of subsystems, in which each subsystem represents all the genes that are involved in a specific pathway [11]. This annotation framework is akin to the one used for the KEGG database, in which proteins are also grouped into distinct pathways [14] and which is seen as an improved method over individual gene annotation because it brings added information about the presence or absence of certain pathways in a metagenome. *IMG/M* [17], another functional annotation pipeline, first processes unannotated sequences through feature prediction packages to identify tRNAs, ribosomal RNAs, protein-coding genes, and CRISPR systems, after which predicted features are functionally annotated through comparison to a combination of pathway databases such as KEGG, and protein family resources such as *HMMER*, *COGs* (cluster of orthologous groups), *Pfam*, and *TIRGfam* [12]. The resulting output is an extensive functional annotation in which predicted proteins are mapped to individual localized cellular functions (e.g. hydrolase activity), as well as to major cellular pathways (e.g. carbon fixation).

It is worth noting that there is a considerable degree of redundancy and complementarity between the different databases and tools used by homology-based pipelines. The level of curation is, in many cases, the major differentiating factor, creating a redundancy in the type of functional data obtained from use of these pipelines. A recent study has shown that different databases offer varying levels of functional prediction and accuracy, with *RefSeq* and *COGs* able to annotate the highest percentage of functional genes from metagenomic sequence data [144]. This variability has also been observed at the pipeline level, with *MG-RAST* being the most reliable at predicting functional shifts in a set of test metagenomic sequence data from different biomes [145].

Recent efforts have been made to develop integrative databases that amalgamate the different individual annotation systems, as is the case with I*nterpro* [145], which combines predictive individual annotations from *PROSITE*, *PRINTS*, *Pfam*, *ProDom*, *SMART*, *TIGRFAMs*, *PIRSF*, and *Gene3D* and is used by the *EBI* metagenomics webserver [16], a European-developed annotation pipeline with comparable performance to *MG-RAST*.

One of the major limitations of homology-based approaches lies in the coverage of the initial metagenomics data. Partial, truncated, and small proteins generated from small contig assemblies are normally not efficiently annotated due to poor similarity with the available database sequences, leading to a large percentage of unannotated sequences [144]. The complexity of the environment from which the metagenomic data are generated can also negatively impact downstream functional analysis due to a high proportion of proteins in those environments that have not yet been documented [144]. To alleviate this issue, some pipelines use context-based approaches that take into account the immediate genomic neighborhood of a sequence to predict function. *SmashCommunity*, another metagenomics analysis tool, combines conventional homology-based approaches for protein annotation with tools that permits an estimation of the quantitative abundances of query sequences in the metagenomes and assembles metagenomic reads into multidomains and operons that give additional functional context to those reads [18].

The bioinformatics pipelines described above are essential in deciphering of the functional capacity of a given environmental metagenomic sequence dataset and have been used to great effect to answer questions around the metabolic lifestyle and survival of microbial communities in extreme environments. In a study performed on Illumina Hiseq-2000 metagenomic data from hypolithic community samples from the Namib desert, a combination of the *MEGAN* pipeline and the *MAPLE* server, which calculates the completeness of a given pathway from the KO terms attributed to specific genes in the metagenomic data, was able to show that members of the Actinobacteria, Proteobacteria, and Cyanobacteria phyla probably play significant roles in primary productivity (photosynthesis, carbon, phosphorus, and sulfur metabolism) [146]. Another metagenomic study on samples from polar desert soils used the *MG-RAST* pipeline to identify genes, present in both biomes, that are involved in the general adaptation of microbial communities to these cold deserts [147]. The study focused on key stress response elements such as chaperone and trehalose synthesis genes and identified biome specific variations in stress-specific genes (such as the alternative sigma factor, sigma B [46]). *MG-RAST* was also used to highlight the presence of nutrient cycling pathways (particularly nitrogen, sulfur, and methane metabolism) and stress response elements in a hypersaline spring microbial community in the High-Arctic [148].

While the bioinformatics platforms used in these studies are still limited by database and raw data constraints (only 20–30% of the total metagenomic sequences can typically be functionally annotated), they already provide detailed information about taxon-related functional capacity and variation in microbial communities, both of which are crucial to a holistic understanding of how these communities adapt and survive in extreme environments.

5.6.2 Application of gene-based microarrays to functional ecology

The development of functional gene arrays (FGAs) has complemented the functional data obtained from metagenomic sequence assemblies by allowing for a direct and rapid high-throughput analysis of microbial communities in environmental samples. Designed in 2004, these environmental FGAs (GeoChips®) operate on the basis of oligonucleotide probes that selectively hybridize with sample sequences of targeted functional genes, generally those involved in geochemical processes such as nitrogen and carbon cycling, metal oxidation and resistance, and organic contaminant degradation. Detection occurs via a fluorescence-based method, providing both qualitative and semi-quantitative data on the functional genes in an environmental sample [149, 150]. For environmental DNA samples, Geochip® hybridization technologies suffer from a number of inherent limitations: the complexity of the samples, the high degree of evolution driven sequence variation in functional genes, and the overcrowding of sequences with unknown function that negatively impacts the sensitivity of the arrays [151]. In recent years, with an increase in the availability of metagenomic data from environmental samples, and our increasing knowledge of the functional roles that microbial communities and their genomic elements play in

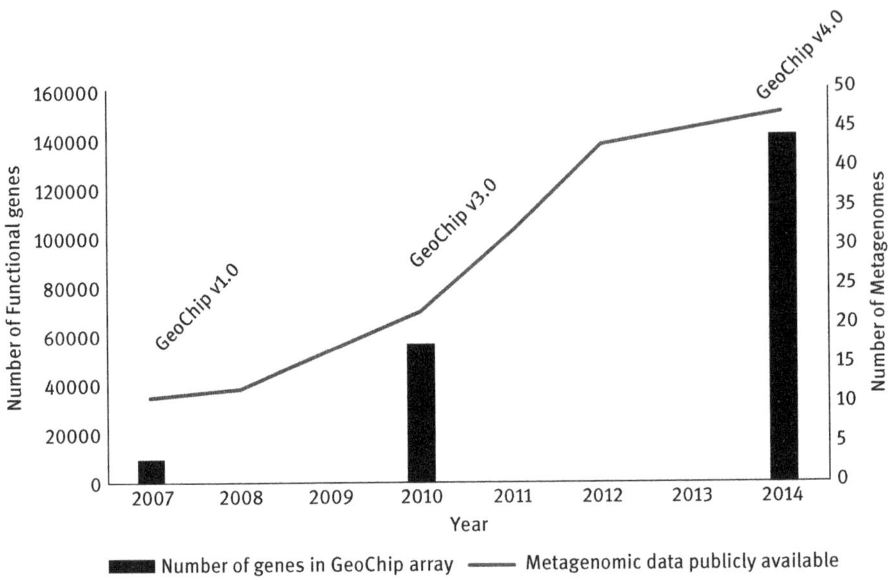

Fig. 5.1: An 8-year comparison of the number of publicly available environmental metagenomes and the gene coverage of GeoChip microarrays. The data on publicly available metagenomes were taken from MG-RAST.

the environment, GeoChip® technology has gone through several iterations to alleviate some of these shortcomings (Fig. 5.1) [151, 152]. In its latest revision, GeoChip® 4.0 allows for the screening of 141,995 sequences from 410 functional genes classes belonging to 12 distinct geochemical pathways across archaeal, bacterial, eukaryotic, and viral taxa [153].

Given their high data yield, it is perhaps not surprising that functional gene microarrays have been used to access the functional capacity and dynamics of microbial communities in various extreme environments. A literature search for reports on the use of GeoChip® assays and related technologies for extremophilic microbial communities shows 10 different studies over the past decade. For example, the functional ecology of microbial communities in the Antarctic Dry Valleys was assessed by combining a comprehensive GeoChip® screen with a 16S PCR-based diversity survey [154]. The authors were able to document the functional processes that are prioritized by communities in this extreme environment, the major microbial taxa that contributed to the functional capacity of these communities, and variations in this functional capacity according to community niche. Similarly, in a GeoChip®-based study focusing on a latitude gradient across Antarctic soils, nutrient cycle-related gene abundance was found to vary according to localization and vegetation, where C- and N-fixation genes were more abundant in rocky and gravel terrain with scarce vegetation, correlating with the need to preserve carbon and nitrogen in these environments [155]. Functional microarrays have also been used to detect the distinct microbial populations that are involved in critical metabolic processes (CO_2 fixation, methane cycling, and nitrogen cycling) in samples from hydrothermal vent chimneys from the Juan de Fuca Ridge [156].

These studies have demonstrated that while the resolving power and accuracy of DNA-hybridization microarrays depend on the input of metagenomic sequencing data as a design criterion, this technology provides an important transition step from annotation and prediction of functional processes to direct quantification of the contribution of these processes to the functional ecology of the microbial communities in extreme environments.

5.7 Applied (functional) metagenomics

Functional metagenomic screening involves the detection of a "functional activity" (such as an enzymic activity, ligand-receptor binding, etc.), independent of any sequence knowledge. The technology of *functional metagenomics* is particularly focused on the discovery of novel enzymes, including new enzyme classes and biocatalysts with unfamiliar reactions. Given our inability to predict extremophilic features (such as thermostability) in annotated protein families *in silico*, functional screening remains the best option for the selection of enzymes with a defined range of biochemical capabilities, such as the ability to operate under harsh reaction conditions processes. Traditional functional screening relies on the construction of nucleic acid libraries containing cloned metagenomic DNA (or cDNA), which are then expressed in a culturable host to identify the targeted activity. However, over the past 2 years, microfluidic-based screening approaches have begun to dominate the field, due to their capacity to screen libraries of unprecedented size with high sensitivity for the identification of rare genes and catalytically versatile enzymes [157]. Irrespective of the approach, functional screening for desired extremophilic properties involves critical decisions regarding the source of the environmental DNA, the type of library to be constructed, the expression host to be selected, and the most viable assay.

5.7.1 Sample selection

Since it is reasonably expected that an enzyme's biochemical properties will reflect the environmental conditions in which it has evolved, the environmental source needs to be selected appropriately. This is most relevant of extreme-temperature, high-pressure, and high-salinity environments, where the intracellular cell environment is at "equilibrium" with the extracellular environment and where both intra- and extracellular biochemical machineries are expected to be adapted to such conditions [158]. In other extreme environments, extremophilic organisms invest considerable metabolic resource to maintain a disequilibrium in cytoplasmic conditions (more typical of non-extremophiles). In consequence, adaptations to extremes of pH, tolerance to organic solvents, or high levels of irradiation are likely only to be found in extracellular enzymes.

Functional metagenomic "bioprospecting" in cold environments, such as polar soils, glacial ice, and deep ocean waters, and hot environments, such as deep sea hydrothermal vents and terrestrial hot springs, has resulted in some of the most successful discoveries (see Refs. [159] and [160] and references therein). Metagenomic screening of acidic and alkaline environments has successfully yielded highly acidophilic and alkaliphic enzymes [159]. However, so-called extremophilic enzymes can be found in unexpected habitats: animal gut microflora have provided an unexpected source of enzymes possessing low pH optima and high tolerances to organic solvents [161].

5.7.2 Metagenomic library construction

Following the identification of an appropriate environment, mDNA is extracted and used for library construction. Specific considerations apply in the preparation of metagenomic expression libraries: both DNA quantity and quality (purity and fragment length) are particularly important. Both direct and indirect extraction methods (see Section 5.2) have application. The advantage of direct extraction is that free DNA from previously lysed organisms (legacy DNA) can also be captured, and mDNA yields tend to be higher than with alternative indirect methods. However, indirect mDNA extraction methods are particularly suited for acidic and alkaliphilic environmental samples, where the sample pH tends to denature or degrade the DNA if extracted directly, and hypersaline samples, where the high concentrations of coextracted salts severely inhibit downstream processes.

Given that most extreme environments have relatively low biomass, DNA yields are often too low for library construction and MDA with φ29 DNA polymerase may be required, either prior to library construction [162] or within picodroplets [163].

The vector to be used for library construction will ultimately dictate the quality of the mDNA required. Plasmid libraries, suitable for single gene products such as enzymes, contain <15 kb inserts and thus can more easily tolerate sheared mDNA preparations. When whole operons and gene clusters are targeted, cosmids/fosmids (25–40 kb) and BACs (100–200 kb) are most appropriate. However, in these instances, the quality of the library is highly sensitive to the mDNA fragment size. The extraction of >40 kb mDNA is crucial but difficult to achieve in practice, particularly from extreme environmental samples.

The discovery of extremozymes from metagenomic libraries prepared from extreme environments has been equally successful using fosmid and plasmid based libraries (see Ref. [160] and references therein), and while the screening of BAC libraries has also been achieved, the average insert size was less than 15 kb [164, 165]. The construction of a metagenomic library is time-consuming and laborious and requires a high level of skill. Depending on the starting material, many different treatments must be trialed in order to successfully generate high-quality large-insert DNA [166, 167].

5.7.3 Expression Hosts

For functional metagenomic screening for "extremozymes," the most suitable expression host would be an extremophile possessing the growth and metabolic requirements that accommodate the properties for which the enzyme/activity of interest is being screened (i.e. a thermophilic host is, at least in theory, the most suitable screening platform for targeting thermophilic enzymes). Considering that significant attention has been devoted to screening extremophilic environmental sample for "extremozymes," the very limited available range of extremophilic hosts for mLibrary construction and screening is perhaps surprising. To date, *Thermus thermophilus* is the only extremophile that has been employed for functional screening of mLibraries [168].

Due to the lack of genetic tools for the majority of culturable bacteria, *Escherichia coli* remains by far the most commonly used host for mLibrary construction and expression. Since functional expression faces the same challenges as does heterologous expression [169], the fraction of extremophilic genes that are successfully expressed is very much lower than represented in a mLibrary (often only a few percentages) [168].

A number of broad-host range cosmid and shuttle vectors have been developed for the parallel screening of mLibraries in diverse proteobacteria, *Bacillus* and *Streptomyces* species ([170] and references therein), to increase mLibraries expression efficiency. However, none of these hosts are suitable to screen for extremophilic properties such as for cold-, alkaline-, and acidic-active enzymes.

5.7.4 Activity assays

Given that the major attraction of extremozymes is enhanced performance under harsh industrial conditions, it is surprising that there are very few examples of enzymes that have been translated into a process [171]. This is in part due to the failure of screening assays to mimic the application settings and/or the industrial process conditions to identify enzymes that meet the industrial criteria, i.e. (i) functionality in high substrate load reactions, (ii) broad temperature and pH ranges, (iii) water-deficient reaction conditions, (iv) very high solvent concentrations, (v) process stability, (vi) high stereoselectivity, and (vii) high turnover rates.

The typical approach employed in most studies, where hits are initially identified through complementation of a given trait [169] followed by a one-by-one biochemical characterization of the positive hits, is a highly inefficient strategy to identify enzymes that meet industrial criteria and within the desired time frame of 3 years for its introduction into the market [171].

Agar plate-based screening of mLibraries is the most simple and common selection methodology employed [172]. However, this approach offers a very limited scope and rarely accommodates the level of stringency required to directly select for the property of interest and the most effective catalyst [173]. Agar plate screening cannot also readily be used for very large libraries (>10^5 independent clones). More integrative approaches can improve the selection process of an industrially relevant enzyme, where the initial screening can query the library for enzyme performance under multiple physical and chemical conditions. This has been most successfully applied for the discovery of novel enzymes with broad substrate spectra using multisubstrate screening approaches [174, 175]. Liquid-based screens, using crude host cell lysates, offer more versatility for screening under real (or close-to-real) process conditions that differ from what is optimal for the host growth [175].

The most recent and exciting development in functional metagenomics is microfluidic-based screening. The technology involves the direct screening of catalytic activities from libraries prepared in water-in-oil droplets, followed by the fluorescent detection of positive events [157, 173]. There are currently two possible versions of this application to metagenomic screening. Either a single cell representing a library member, or a single library clone molecule, is co-compartmentalized with a fluorogenic substrate for the target enzyme, along with either the lysis reagents or an *in vitro* transcription/translation reaction mixture, all within a single droplet. In both scenarios, proteins with an appropriate catalytic activity release a fluorescent product. "Positive" microdroplets are separated from non-fluorescent microdroplets using a microfluidic platform, such as a fluorescence activated cell sorter. Vector DNA from the selected microdroplets is recovered, amplified using MDA, sequenced, and analyzed using standard bioinformatic methods for the identification of the gene encoding the detected activity.

Microdroplets can be screened in an ultra-high-throughput manner, at up to 50,000 clones per second or over 1 billion clones per day, a processing speed that is impossible to achieve using current conventional high-throughput screening approaches, including robotic liquid handling platforms [173]. This not only makes accessing the full metagenomic potential of a given sample realistic, but also the method is highly adaptable for screening a variety of biochemical parameters in a relatively short time. Microfluidic methods have the added advantage of fine control of the reaction conditions and the capacity for the introduction of novel chemical entities to enzymes (e.g. new metals, cofactors, and unnatural amino acids).

In addition to their speed, the sensitivity of such microfluidic systems has been shown to facilitate the identification of weak side activities with orders-of-magnitude slower rates than the native activities, a valuable approach for the identification of new and rare enzymes [157]. Although not yet applied to metagenomic screening, stand-alone *in vitro* transcription/translation systems would allow for selection of substrates, products, and reaction conditions that are incompatible with *in vivo* expression systems and are thus particularly attractive for screening under extremophilic conditions. Although an *in vitro* approach is limited by a reliance on, and limited availability of, cell-free systems comprising the essential components for transcription and translation, we predict that in the near future, new *in vitro*-based expression platforms will emerge that will be more amenable to the bioprospecting of extremozymes and that could significantly influence the number of extremophilic enzymes and activities finding industrial application.

5.8 Conclusions and perspectives

The advent and democratization of high-throughput sequencing technologies that allow for the study of population genomics have led to an explosion of sequence-based information that can be accessed in the public scientific space. This ever-expanding database of information needs accurate functional and taxonomical curation, and the crucial role of bioinformatic tools in this curation process cannot be understated. Robust annotation pipelines such as MG-RAST and MEGAN support the study of large metagenomic datasets at a high resolution, often allowing for rationalization of data to genus level and or even the level of individual metabolic pathways. However, these pipelines are multi-algorithm methods that require a significant investment of computational resources in order to parse large metagenomic datasets. They are also limited by the level of curation of the multiple sequence databases they employ and, correspondingly, the level of empirical knowledge that we have accrued over the decades. This limitation, in particular, makes the bioinformatics tools used for metagenomics biased toward better studied and understood environments.

Both targeted and whole metagenome sequencing of DNA provide snapshots of the gene content and genetic diversity of microbial communities. However, there are important differences between these approaches in terms of sample preparation, sequence output, data analysis, and economic cost. For example, targeted sequencing can provide greater depth of coverage for specific gene(s) of interest (e.g. 16S rRNA and nifH), while whole metagenome sequencing captures information about the entire genomic content of a sample, providing the opportunity to simultaneously explore the taxonomic and functional diversity of microbial communities. Therefore, it is important to keep in mind that each technology has its strengths and weaknesses and must be selected based on the biological questions and objectives of the study.

In the same context, many microbiome studies are geared toward the discovery and description of the prokaryotic component of the microbiome, and in the process, the virome is largely overlooked. There is no easy way to incorporate a viral component into a standard phylogenetic pipeline, given the lack of a signature gene for amplicon studies and because of the discrepancy in genome sizes between most viruses and their hosts. Necessarily, virome studies must be performed in parallel with microbiomic surveys, using separate enrichment and amplification protocols. This requirement represents a practical weakness in current microbial ecology studies (although the same could be said for ecological studies on other taxa, such as microeukaryotes, microfauna, protists, etc.).

It is also worth considering the question as to whether, in this era of "mega-sequencing," functional metagenomics is now redundant technology. In a search for new gene products and functionalities, particularly in the fields of applied microbiology, biocatalysis, and industrial enzymology, can one acquire any target from the terrabase pairs of nucleic acid sequence that are already publically available in metagenomic sequence databases (and which are growing exponentially)?

Gene mining from metagenomic sequence data currently lacks the functional specificity information potentially available through the screening methods of functional metagenomics. While current gene annotation technologies can (mostly) provide reliable information on the protein category and, therefore, the functional class of an enzyme for example, the resolution of these bioinformatic processes is currently completely incapable of reliably yielding information on general protein properties (e.g. thermostability), let alone more high-resolution functional parameters such as substrate or ligand binding affinity or enzyme catalytic properties such as regio- or stereo-selectivity or turnover number.

"Looking back in order to see forward," one might also predict that the relative value of both metagenome sequencing and functional metagenomics will change substantially, probably within the next decade. The development of more complex and sophisticated predictive algorithms for deriving structure, and function, from sequence are very likely to steadily shift the balance towards direct sequence data mining. Recent reports [e.g. 176–178] are testaments of the rapid progress in sequence-to-function predictive technologies!

The future time when the bioinformatic toolkits are sufficiently sophisticated for a complete pipeline analysis, from sequence to detailed function, may be less distant than we think!

References

[1] Venter JC, Remington K, Heidelberg JF, et al. Environmental genome shotgun sequencing of the Sargasso Sea. Science. 2004;304:66–74.

[2] Tyson GW, Chapman J, Hugenholtz P, et al. Community structure and metabolism through reconstruction of microbial genomes from the environment. Nature. 2004;428:37–43.

[3] Anantharaman K, Brown CT, Hug LA, et al. Thousands of microbial genomes shed light on interconnected biogeochemical processes in an aquifer system. Nature Commun. 2016;7:13219.
[4] Delmont TO, Robe P, Clark I, Simonet P, Vogel TM. Metagenomic comparison of direct and indirect soil DNA extraction approaches. J Microbiol Meth. 2011;86:397–400.
[5] Herrera A, Cockell CS. Exploring microbial diversity in volcanic environments: a review of methods in DNA extraction. J Microbiol Meth. 2007;70:1–12.
[6] Bertin PN, Heinrich-Salmeron A, Pelletier E, et al. Metabolic diversity among main microorganisms inside an arsenic-rich ecosystem revealed by meta- and proteo-genomics. ISME J. 2011;5:1735–47.
[7] Tighe S, Afshinnekoo E, Rock TM, et al. Genomic methods and microbiological technologies for profiling novel and extreme environments for the Extreme Microbiome Project (XMP). J Biomol Tech. 2017;28:31.
[8] Bond PL, Smriga SP, Banfield JF. Phylogeny of microorganisms populating a thick, subaerial, predominantly lithotrophic biofilm at an extreme acid mine drainage site. Appl Environ Microbiol. 2000;66:3842–9.
[9] Čanković M, Petrić I, Marguš M, Ciglenečki, I. Spatio-temporal dynamics of sulfate-reducing bacteria in extreme environment of Rogoznica Lake revealed by 16S rRNA analysis. J Marine Syst. 2017;172:14–23.
[10] Meyer FD, Paarmann M, D'Souza R, et al. The metagenomics RAST server – a public resource for the automatic phylogenetic and functional analysis of metagenomes. BMC Bioinform. 2008;9:386.
[11] Overbeek R, Begley T, Butler RM, et al. The subsystems approach to genome annotation and its use in the project to annotate 1000 genomes. Nucleic Acids Res. 2005;33:5691–702.
[12] Chen IA, Markowitz VM, Chu K, et al. IMG/M: integrated genome and metagenome comparative data analysis system. Nucleic Acids Res. 2017;45:D507–16.
[13] Sayers EW, Barrett T, Benson DA, et al. Database resources of the National Center for Biotechnology Information. Nucleic Acids Res. 2012;40:D13–25.
[14] Kanehisa M, Goto S. KEGG: Kyoto encyclopedia of genes and genomes. Nucleic Acids Res. 2000;28;27–30.
[15] Tatusov RL, Koonin EV, Lipman DJ. A genomic perspective on protein families. Science. 1997;278:631–7.
[16] Hunter S, Apweiler R, Attwood TK, et al. InterPro: the integrative protein signature database. Nucleic Acids Res. 2009;37:D211–5.
[17] Markowitz VM, Chen IM, Palaniappan K, et al. IMG: the integrated microbial genomes database and comparative analysis system. Nucleic Acid Res. 2012;40:D115–22.
[18] Hunter S, Corbett M, Denise H, et al. EBI metagenomics – a new resource for the analysis and archiving of metagenomics data. Nucleic Acids Res. 2014;42:D600–6.
[19] Lees J, Yeats C, Perkins J, et al. Gene3D: a domain-based resource for comparative genomics, functional annotation and protein network analysis. Nucleic Acid Res. 2012;40:D465–71.
[20] Attwood TK, Bradley P, Flower DR, et al. PRINTS and its automatic supplement, prePRINTS. Nucleic Acids Res. 2003;31:400–2.
[21] Punta M, Coggill PC, Eberhardt RY et al. The Pfam protein families database. Nucleic Acids Res. 2012;40:D290–301.
[22] Selengut JD, Haft DH, Davidsen T, et al. TIGRFAMs and genome properties: tools for the assignment of molecular function and biological process in prokaryotic genomes. Nucleic Acids Res. 2007;35:D260–4.
[23] Sigrist CJ, Cerutti L, de Castro E, et al. PROSITE, a protein domain database for functional characterization and annotation. Nucleic Acids Res. 2010;38:D161–6.

[24] Ashburner M, Ball CA, Blake JA et al. Gene Ontology: tool for the unification of biology, Nature Genet. 2000;25:25–9.
[25] Wu S, Zhu Z, Fu L, Niu B, Li W. WebMGA: a customizable web server for fast metagenomics sequence analysis. BMC Genom. 2011;12:444.
[26] Huson DH, Auch AF, Qi J, Schuster SC. MEGAN analysis of metagenomics data. Genome Res. 2007;17:377–86.
[27] Gabor EM, Vries EJ, Janssen DB. Efficient recovery of environmental DNA for expression cloning by indirect extraction methods. FEMS Microbiol Ecol. 2003;44:153–63.
[28] Poté J, Bravo AG, Mavingui P, Ariztegui D, Wildi W. Evaluation of quantitative recovery of bacterial cells and DNA from different lake sediments by Nycodenz density gradient centrifugation. Ecological Indic. 2010;10:234–40.
[29] Jacobsen CS, Rasmussen OF. Development and application of a new method to extract bacterial DNA from soil based on separation of bacteria from soil with cation-exchange resin. Appl Environ Microbiol. 1992;58:2458–62.
[30] Hedlund BP, Dodsworth JA, Murugapiran SK, Rinke C, Woyke T. Impact of single-cell genomics and metagenomics on the emerging view of extremophile "microbial dark matter". Extremophiles. 2014;18:865–75.
[31] Bokulich NA, Subramanian S, Faith JJ, et al. Quality-filtering vastly improves diversity estimates from Illumina amplicon sequencing. Nature Meth. 2013;10:57–9.
[32] Schloss PD, Handelsman J. Status of the microbial census. Microbiol Molec Biol Rev. 2004;68:686–91.
[33] Westcott SL, Schloss PD. De novo clustering methods outperform reference-based methods for assigning 16S rRNA gene sequences to operational taxonomic units. Peer J. 2015;3:e148.
[34] Rideout JR, He Y, Navas-Molina JA, et al. Subsampled open-reference clustering creates consistent, comprehensive OTU definitions and scales to billions of sequences. Peer J. 2014;2;e545.
[35] Schloss PD, Westcott SL, Ryabin T, et al. Introducing mothur: open-source, platform-independent, community-supported software for describing and comparing microbial communities. Appl Environ Microbiol. 2009;75:7537–41.
[36] Edgar RC. Search and clustering orders of magnitude faster than BLAST. Bioinformatics. 2010;26:2460–1.
[37] Rognes T, Flouri T, Nichols B, Quince C, Mahé F. VSEARCH: a versatile open source tool for metagenomics. Peer J. 2016;4:e2584.
[38] Haas BJ, Gevers D, Earl AM, et al. Chimeric 16S rRNA sequence formation and detection in Sanger and 454-pyrosequenced PCR amplicons. Genom Res. 2011;21;494–504.
[39] Edgar RC, Haas BJ, Clemente JC, Quince C, Knight R. UCHIME improves sensitivity and speed of chimera detection. Bioinformatics. 2011;27:2194–200.
[40] Cole JR, Wang Q, Cardenas E, et al. The Ribosomal Database Project: improved alignments and new tools for rRNA analysis. Nucleic Acids Res. 2009;37:D141–D5.
[41] Quast C, Pruesse E, Yilmaz P, et al. The SILVA ribosomal RNA gene database project: improved data processing and web-based tools. Nucl Acids Res. 2013;41:D590–6.
[42] DeSantis TZ, Hugenholtz P, Larsen N, et al. Greengenes, a chimera-checked 16S rRNA gene database and workbench compatible with ARB. Appl Environ Microbiol. 2006;72:5069–72.
[43] Caporaso JG, Bittinger K, Bushman FD, DeSantis TZ, Andersen GL, Knight R. PyNAST: a flexible tool for aligning sequences to a template alignment. Bioinformatics. 2010;26:266–7.
[44] Larkin MA, Blackshields G, Brown NP, et al. Clustal W and Clustal X version 2.0. Bioinformatics. 2007;23:2947–8.
[45] Edgar RC. MUSCLE: multiple sequence alignment with high accuracy and high throughput. Nucl Acids Res. 2004;32:1792–7.

[46] Price MN, Dehal PS, Arkin AP. FastTree 2-approximately maximum-likelihood trees for large alignments. PloS One. 2010;5.
[47] Stamatakis A. RAxML-VI-HPC: Maximum likelihood-based phylogenetic analyses with thousands of taxa and mixed models. Bioinformatics. 2006;22:2688–90.
[48] McMurdie PJ, Holmes S. Waste not, want not: why rarefying microbiome data is inadmissible. PLoS Comput Biol. 2014;10:e1003531.
[49] Tiao G, Lee CK, McDonald IR, Cowan DA, Cary SC. Rapid microbial response to the presence of an ancient relic in the Antarctic Dry Valleys. Nature Commun. 2012;3:360.
[50] Magurran AE. Measuring biological diversity. John Wiley & Sons; 2004. 256 pp.
[51] Faith DP. Conservation evaluation and phylogenetic diversity. Biol Conservat. 1992;61:1–10.
[52] Fierer N, Leff JW, Adams BJ, et al. Cross-biome metagenomic analyses of soil microbial communities and their functional attributes. Proc Natl Acad Sci USA. 2012;109:21390–5.
[53] Sharp CE, Brady AL, Sharp GH, Grasby SE, Stott MB, Dunfield PF. Humboldt's spa: microbial diversity is controlled by temperature in geothermal environments. ISME J. 2014;8:1166–74.
[54] Anderson MJ, Crist TO, Chase JM, et al. Navigating the multiple meanings of beta-diversity: a roadmap for the practicing ecologist. Ecology Lett. 2011;14:19–28.
[55] Lozupone C, Knight R. UniFrac: A new phylogenetic method for comparing microbial communities. Appl Environ Microbiol. 2005;71:8228–35.
[56] Kuczynski J, Liu Z, Lozupone C, McDonald D, Fierer N, Knight R. Microbial community resemblance methods differ in their ability to detect biologically relevant patterns. Nature Meth. 2010;7:813–U67.
[57] Mackelprang R, Waldrop MP, DeAngelis KM, et al. Metagenomic analysis of a permafrost microbial community reveals a rapid response to thaw. Nature. 2011;480:368–71.
[58] Liu J, Hua ZS, Chen LX, et al. Correlating microbial diversity patterns with geochemistry in an extreme and heterogeneous environment of mine tailings. Appl Environ Microbiol. 2014;80:3677–86.
[59] Mantel N. The detection of disease clustering and a generalized regression approach. Cancer Res. 1967;27:209–20.
[60] Clarke KR. Non-parametric multivariate analyses of changes in community structure. Austral Ecol. 1993;18:117–43.
[61] Anderson MJ. A new method for non-parametric multivariate analysis of variance. Aust Ecol. 2001;26:32–46.
[62] Dufrene M, Legendre P. Species assemblages and indicator species: the need for a flexible asymmetrical approach. Ecolog Monogr. 1997;67:345–66.
[63] Segata N, Izard J, Waldron L, et al. Metagenomic biomarker discovery and explanation. Genome Biol. 2011;12:R60.
[64] Makhalanyane TP, Valverde A, Lacap DC, Pointing SB, Tuffin MI, Cowan DA. Evidence of species recruitment and development of hot desert hypolithic communities. Environ Microbiol Rep 2013;5:219–24.
[65] Caporaso JG, Kuczynski J, Stombaugh J, et al. QIIME allows analysis of high-throughput community sequencing data. Nature Meth. 2010;7:335–6.
[66] Oksanen J, Blanchet FG, Kindt R, et al. Vegan: Community Ecology Package. R package version 2.0-2. Accessed July 24, 2017, at https://cran.r-project.org/web/packages/vegan/index.html.
[67] Dray S, Dufour AB. The ade4 package: implementing the duality diagram for ecologists. J Statist Software. 2007;22:1–20.
[68] Paradis E, Claude J, Strimmer K. APE: analyses of phylogenetics and evolution in R language. Bioinformatics. 2004;20:289–90.
[69] McMurdie PJ, Holmes S. Phyloseq: an R package for reproducible interactive analysis and graphics of microbiome census data. PLoS ONE. 2013;8:e61217.

[70] Adriaenssens EM, Cowan DA. Using signature genes as tools to assess environmental viral ecology and diversity. Appl Environ Microbiol. 2014;80:4470–80.
[71] Rohwer F, Edwards R. The Phage Proteomic Tree: a genome-based taxonomy for phage. J Bacteriol. 2002;184:4529–35.
[72] Vega Thurber R, Haynes M, Breitbart M, Wegley L, Rohwer F. Laboratory procedures to generate viral metagenomes. Nature Protoc. 2009;4:470–83.
[73] Adriaenssens EM, van Zyl L, Cowan D, Trindade M. Metaviromics of Namib Desert salt sans: A novel lineage of haloarchaeal Salterproviruses and a rich source of ssDNA viruses. Viruses. 2016;8:14.
[74] Diemer GS, Stedman KM. A novel virus genome discovered in an extreme environment suggests recombination between unrelated groups of RNA and DNA viruses. Biol Direct. 2012;7:13.
[75] Emerson JB, Thomas BC, Andrade K, Allen EE, Heidelberg KB, Banfield JF. Dynamic viral populations in hypersaline systems as revealed by metagenomic assembly. Appl Environ Microbiol. 2012;78:6309–20.
[76] Garcia-Heredia I, Martin-Cuadrado AB, Mojica FJM, et al. Reconstructing viral genomes from the environment using fosmid clones: the case of haloviruses. PLoS One. 2012;7:e33802.
[77] Santos F, Meyerdierks A, Peña A, Rosselló-Mora R, Amann R, Antón J. Metagenomic approach to the study of halophages: the environmental halophage 1. Environ Microbiol. 2007;9:1711–23.
[78] Santos F, Yarza P, Parro V, Briones C, Anton J. The metavirome of a hypersaline environment. Environ Microbiol. 2010;12:2965–76.
[79] Schoenfeld T, Patterson M, Richardson PM, Wommack KE, Young M, Mead D. Assembly of viral metagenomes from Yellowstone hot springs. Appl Environ Microbiol. 2008;74:4164–74.
[80] Bolduc B, Shaughnessy DP, Wolf YI, Koonin EV, Roberto FF, Young M. Identification of novel positive-strand RNA viruses by metagenomic analysis of archaea-dominated Yellowstone hot springs. J Virol. 2012;86:5562–73.
[81] Hurwitz BL, Deng L, Poulos BT, Sullivan MB. Evaluation of methods to concentrate and purify ocean virus communities through comparative, replicated metagenomics. Environ Microbiol. 2013;15:1428–40.
[82] John SG, Mendez CB, Deng L, et al. A simple and efficient method for concentration of ocean viruses by chemical flocculation. Environ Microbiol Rep. 2011;3:195–202.
[83] Williamson KE, Radosevich M, Wommack KE. Abundance and diversity of viruses in six Delaware soils. Appl Environ Microbiol. 2005;71:3119–25.
[84] Williamson KE, Wommack KE, Radosevich M. Sampling natural viral communities from soil for culture-independent analyses. Appl Environ Microbiol. 2003;69:6628–33.
[85] Zablocki O, Adriaenssens EM, Frossard A, Seely M, Ramond J-B, Cowan D. Metaviromes of extracellular soil viruses along a Namib desert aridity gradient. Genome Announc. 2017;5:e01470–16.
[86] Adriaenssens EM, Van Zyl L, De Maayer P, et al. Metagenomic analysis of the viral community in Namib Desert hypoliths. Environ Microbiol. 2015;17:480–95.
[87] Hesse U, van Heusden P, Kirby BM, Olonade I, van Zyl LJ, Trindade M. Virome assembly and annotation: a surprise in the Namib Desert. Front Microbiol. 2017;8:1–17.
[88] Zablocki O, Zyl L Van, Adriaenssens EM, et al. High-level diversity of tailed phages, eukaryote-associated viruses and virophage-like elements in the metaviromes of Antarctic soils. Appl Environ Microbiol. 2014;80:6888–97.
[89] Fierer N, Breitbart M, Nulton J, et al. Metagenomic and small-subunit rRNA analyses reveal the genetic diversity of bacteria, archaea, fungi, and viruses in soil. Appl Environ Microbiol. 2007;73,:7059–66.
[90] Yoshida M, Takaki Y, Eitoku M, Nunoura T, Takai K. Metagenomic analysis of viral communities in (hado)pelagic sediments. PLoS One. 2008;8:e57271.

[91] Appelt S, Fancello L, Le Bailly M, Raoult D, Drancourt M, Desnues C. Viruses in a 14th-century coprolite. Appl Environ Microbiol. 2014;80:2648–55.
[92] Sime-Ngando T, Lucas S, Robin A, et al. Diversity of virus-host systems in hypersaline Lake Retba, Senegal. Environ Microbiol. 2011;13:1956–72.
[93] Hall RJ, Wang J, Todd AK, et al. Evaluation of rapid and simple techniques for the enrichment of viruses prior to metagenomic virus discovery. J Virol Methods. 2014;195:194–204.
[94] Woo PCY, Lau SKP, Teng JLL, et al. Metagenomic analysis of viromes of dromedary camel fecal samples reveals large number and high diversity of circoviruses and picobirnaviruses. Virology. 2014;471:117–25.
[95] Daly GM, Bexfield N, Heaney J, et al. A viral discovery methodology for clinical biopsy samples utilising massively parallel next generation sequencing. PLoS One. 2011;6:e28879.
[96] Kim K-H, Bae J-W. Amplification methods bias metagenomic libraries of uncultured single-stranded and double-stranded DNA viruses. Appl Environ Microbiol. 2011;77:7663–8.
[97] Marine R, McCarren C, Vorrasane V, et al. Caught in the middle with multiple displacement amplification: the myth of pooling for avoiding multiple displacement amplification bias in a metagenome. Microbiome. 2014;2:3.
[98] Polson SW, Wilhelm SW, Wommack KE. Unraveling the viral tapestry (from inside the capsid out). ISME J. 2011;5:165–8.
[99] Djikeng A, Halpin R, Kuzmickas R, et al. Viral genome sequencing by random priming methods. BMC Genomics. 2008;9:5.
[100] Karlsson OE, Belák S, Granberg F. The effect of preprocessing by sequence-independent, single-primer amplification (SISPA) on metagenomic detection of viruses. Biosec Bioterror Biodef Strateg Pract Sci. 2013;11:S227–34.
[101] Miranda JA, Culley AI, Schvarcz CR, Steward GF. RNA viruses as major contributors to Antarctic virioplankton. Environ Microbiol. 2016;18:3714–27.
[102] Weynberg KD, Wood-Charlson EM, Suttle CA, van Oppen MJH. Generating viral metagenomes from the coral holobiont. Front Microbiol. 2014;5:e206.
[103] Prosser JI. Dispersing misconceptions and identifying opportunities for the use of 'omics' in soil microbial ecology. Nat Rev Microbiol. 2015;13:439–46.
[104] Andrews S. FastQC: a quality control tool for high throughput sequence data. 2010. Accessed June 9, 2017, at http://www.bioinformatics.babraham.ac.uk/projects/fastqc.
[105] Bolger AM, Lohse M, Usadel B. Trimmomatic: a flexible trimmer for Illumina sequence data. Bioinformatics. 2014;btu170.
[106] Patel RK, Jain M. NGS QC Toolkit: a toolkit for quality control of next generation sequencing data. PloS One. 2012;7:e30619.
[107] Schmieder R, Edwards R. Quality control and preprocessing of metagenomic datasets. Bioinformatics. 2011;27:863–4.
[108] Schubert M, Lindgreen S, Orlando L. AdapterRemoval v2: rapid adapter trimming, identification, and read merging. BMC Res Notes. 2016;9:88.
[109] Kunin V, Copeland A, Lapidus A, Mavromatis K, Hugenholtz P. A bioinformatician's guide to metagenomics. Microbiol Molec Biol Rev. 2008;72:557–78.
[110] Gevers D, Pop M, Schloss PD, Huttenhower C. Bioinformatics for the Human Microbiome Project. PLoS Comput Biol. 2012;8:e1002779.
[111] Nurk S, Meleshko D, Korobeynikov A, Pevzner PA. metaSPAdes: a new versatile metagenomic assembler. Genome Res. 2017;27:824–34.
[112] Li D, Luo R, Liu CM, et al. MEGAHIT v1.0: a fast and scalable metagenome assembler driven by advanced methodologies and community practices. Methods. 2016;102:3–11.
[113] Peng Y, Leung H, Yiu SM, Chin FY. IDBA-UD: a de novo assembler for single-cell and metagenomic sequencing data with highly uneven depth. Bioinformatics. 2012;28:1420–8.

[114] Boisvert S, Raymond F, Godzaridis É, Laviolette F, Corbeil J. Ray Meta: scalable de novo metagenome assembly and profiling. Genome Biol. 2012;13:R122.
[115] Van der Walt AJ, Van Goethem MW, Ramond JB, Makhalanyane TP, Reva O, Cowan DA. Assembling metagenomes, one community at a time. BMC Genomics. 2017;18:521.
[116] Vollmers J, Wiegand S, Kaster AK. Comparing and evaluating metagenome assembly tools from a microbiologist's perspective – not only size matters! PLoS One. 2017;12:e0169662.
[117] Mikheenko A, Saveliev V, Gurevich A. MetaQUAST: evaluation of metagenome assemblies. Bioinformatics. 2016;32:1088–90.
[118] Sangwan N, Xia F, Gilbert JA. Recovering complete and draft population genomes from metagenome datasets. Microbiome. 2016;4:8.
[119] Dick GJ, Andersson AF, Baker BJ, et al. Community-wide analysis of microbial genome sequence signatures. Genome Biol. 2009;10:R85.
[120] Teeling H, Waldmann J, Lombardot T. Bauer M, Glöckner FO. TETRA: a web-service and a stand-alone program for the analysis and comparison of tetranucleotide usage patterns in DNA sequences. BMC Bioinform. 2004;5:163.
[121] Wang Y, Leung HC, Yiu SM, Chin FY. MetaCluster 5.0, a two-round binning approach for metagenomic data for low-abundance species in a noisy sample. Bioinformatics. 2012;28:i356–62.
[122] Ghai R, Mizuno CM, Picazo A, Camacho A, Rodriguez-Valera F. Key roles for freshwater Actinobacteria revealed by deep metagenomic sequencing. Molec Ecol. 2014;23:6073–90.
[123] Gibbons SM, Schwartz T, Fouquier J, et al. Ecological succession and viability of human-associated microbiota on restroom surfaces. Appl Environ Microbiol. 2015;81:765–73.
[124] Le Chatelier E, Nielsen T, Qin J, et al. Richness of human gut microbiome correlates with metabolic markers. Nature. 2013;500:541–6.
[125] Wu YW, Ye Y. A novel abundance-based algorithm for binning metagenomic sequences using l-tuples. J Comput Biol. 2011;18:523–34.
[126] Albertsen M, Hugenholtz P, Skarshewski A, Nielsen KL, Tyson GW, Nielsen PH. Genome sequences of rare, uncultured bacteria obtained by differential coverage binning of multiple metagenomes. Nature Biotechnol. 2013;31:533–8.
[127] Wrighton KC, Thomas BC, Sharon I, et al. Fermentation, hydrogen, and sulfur metabolism in multiple uncultivated bacterial phyla. Science. 2012;337:1661–5.
[128] Alneberg J, Bjarnason BS, De Bruijn I. et al, Binning metagenomic contigs by coverage and composition. Nat Methods. 2014;11:1144–6.
[129] Imelfort M, Parks D, Woodcroft BJ, Dennis P, Hugenholtz P, Tyson GW. GroopM: an automated tool for the recovery of population genomes from related metagenomes. Peer J. 2014;2:e603.
[130] Kang DD, Froula J, Egan R, Wang Z. MetaBAT, an efficient tool for accurately reconstructing single genomes from complex microbial communities. Peer J. 2015;3:e1165.
[131] Wu YW, Simmons BA, Singer SW. MaxBin 2.0: an automated binning algorithm to recover genomes from multiple metagenomic datasets. Bioinformatics. 2016;32:605–7.
[132] Sieber CMK, Probst AJ, Sharrar A, et al. Recovery of genomes from metagenomes via a dereplication, aggregation, and scoring strategy. bioRxiv. 2017;107789.
[133] Parks DH, Imelfort M, Skennerton CT, Hugenholtz P, Tyson GW. CheckM: assessing the quality of microbial genomes recovered from isolates, single cells, and metagenomes. Genome Res. 2015;25:1043–55.
[134] Campbell JH, O'Donoghue P, Campbell AG. UGA is an additional glycine codon in uncultured SR1 bacteria from the human microbiota. Proc Natl Acad Sci USA. 2013;110:5540–5.
[135] Rinke C, Schwientek P, Sczyrba A, et al. Insights into the phylogeny and coding potential of microbial dark matter. Nature. 2013;499:431–7.
[136] Narasingarao P, Podell S, Ugalde JA, et al. De novo metagenomic assembly reveals abundant novel major lineage of Archaea in hypersaline microbial communities. ISME J. 2012;6:81–93.

[137] Buongiorno J, Bird JT, Krivushin K, et al. Draft genome sequence of Antarctic methanogen enriched from Dry Valley permafrost. Genome Announc. 2016;4(6):e01362–16.
[138] Johnston ER, Rodriguez-R L.M, Luo C, et al. Metagenomics reveals pervasive bacterial populations and reduced community diversity across the Alaska tundra ecosystem. Front Microbiol. 2016;7:579.
[139] Li M, Jain S, Dick GJ. Genomic and transcriptomic resolution of organic matter utilization among deep-sea bacteria in Guaymas Basin hydrothermal plumes. Front Microbiol. 2016;7:1125.
[140] Haro-Moreno JM, Rodriguez-Valera F, López-García P, Moreira D, Martin-Cuadrado AB. New insights into marine group III Euryarchaeota, from dark to light. ISME J. 2017;11:1102–17.
[141] Skennerton CT, Ward LM, Michel A, Metcalfe K, Valiente C, Mullin S, Chan KY, Gradinaru V, Orphan VJ. Genomic reconstruction of an uncultured hydrothermal vent gammaproteobacterial methanotroph (family Methylothermaceae) indicates multiple adaptations to oxygen limitation. Front Microbiol. 2015;6:1425.
[142] Handelsman J. Metagenomics: Application of genomics to uncultured microorganisms. Microbiol Molec Biol Rev. 2004;68:669–85.
[143] Sharpton TJ. An introduction to the analysis of shotgun metagenomic data. Front Plant Sci. 2014;5:209.
[144] Simon C, Daniel R. Metagenomic analyses: past and future trends. Appl Environ Microbiol. 2005;77:1153–61.
[145] Prakash T, Taylor DT. Functional assignment of metagenomics data: challenges and applications, Brief Bioinform. 2012;13:711–27.
[146] Lindgreen S, Adair KL, Gardner PP. An evaluation of the accuracy and speed of metagenome analysis tools. Scientif Rep. 2016;6:19233.
[147] Vikram, S, Guerrero LD, Makhalanyane TP, Le PT, Seely M, Cowan DA. Metagenomic analysis provides insights into functional capacity in a hyperarid desert soil niche community. Environ Microbiol. 2016;18:1875–88.
[148] Varin T, Lovejoy C, Jungblut AD, Vincent WF, Corbell J. Metagenomic analysis of stress genes in microbial mat communities from Antarctica and the high Arctic. Appl Environ Microbiol. 2012;78:549–59.
[149] Lay CY, Mykytczuk NC, Yergeau É, Lamarche-Gagnon G, Greer GW, Whyte LG. Defining the functional potential and active community members of a sediment microbial community in a high-arctic hypersaline subzero spring. Appl Environ Microbiol. 2013;79:3637–48.
[150] He Z, Gentry TJ, Schadt WC, Wu L et al. GeoChip: A comprehensive microarray for investigating biochemical, ecological, and environmental processes. ISME J. 2007;1:67–77.
[151] Schadt CW, Liebich J, Chong SC, et al. Design and use of Functional Gene Microarrays (FGAs) for the characterization of microbial communities, Meth Microbiol. 2004;34:331–68.
[152] He Z, van Nostrand JD, Wu LY, Zhou JZ. Development and application of functional gene arrays for microbial community analysis. Trans Nonferrous Met Soc China. 2008;18:1319–27.
[153] He Z, Deng Y, Van Nostrand J D et al. GeoChip 3.0 as a high-throughput tool for analysing microbial community composition, structure and functional activity. ISME J. 2010;4:1167–79.
[154] Tu Q, Yu H, He Z et al. GeoChip 4: a functional gene-array-based high throughput environmental technology for microbial community analysis. Molec Ecol Res. 2014;14:914–28.
[155] Chan Y, van Nostrand JD, Zhou J, Pointing SB, Farrell L. Functional ecology of an Antarctic Dry Valley. Proc Natl Acad Sci USA. 2013;110:8990–5.
[156] Yergeau E, Kang S, He Z, Zhou J, Kowalchuk GA. Functional microarray analysis of nitrogen and carbon cycling genes across an Antarctic latitudinal transect. ISME J. 2007;1:163–79.
[157] Wang F, Zhou H, Meng J et al. GeoChip-based analysis of metabolic diversity of microbial communities at the Juan de Fuca Ridge hydrothermal vent. Proc Natl Acad Sci USA. 2009;106:4840–5.

[158] Colin P-Y, Kintses B, Gielen F, et al. Ultrahigh-throughput discovery of promiscuous enzymes by picodroplet functional metagenomics. Nature Commun. 2015;6:10008.
[159] Ferrer M, Golyshina O, Beloqui A, Golyshin P. Mining enzymes from extreme environments. Curr Opin Microbiol. 2007;10:207–14.
[160] Mirete S, Morgante V, Gonzalez-Pastor J. Functional metagenomics of extreme environments. Curr Opin Biotechnol. 2016;38:143–9.
[161] Ferrer M, Golyshina OV, Chernikova TN, et al. Novel hydrolase diversity retrieved from a metagenome library of bovine rumen microflora. Environ Microbiol. 2005;7:1996–2010.
[162] Spits C, Le Caignec C, De Rycke M, et al. Whole-genome multiple displacement amplification from single cells. Nature Protoc. 2006;1:1965–70.
[163] Hammond M, Homa F, Andersson-svahn H, Ettema T, Joensson H. Picodroplet partitioned whole genome amplification of low biomass samples preserves genomic diversity for metagenomic analysis. Microbiome. 2016;4:52.
[164] Berlemont R, Delsaute M, Pipers D, et al. Insights into bacterial cellulose biosynthesis by functional metagenomics on Antarctic soil samples. ISME J. 2009;3:1070–81.
[165] Vester JK, Glaring MA, Stougaard P. Discovery of novel enzymes with industrial potential from a cold and alkaline environment by a combination of functional metagenomics and culturing. Microb Cell Fact. 2014;13:72.
[166] Lam K, Cheng J, Engel K, Neufeld J, Charles T. Current and future resources for functional metagenomics. Front Microbiol. 2015;6:1196.
[167] Liles M, Williamson L, Rodbumrer J, Torsvik V, Goodman R, Handelsman J. Recovery, purification, and cloning of high-molecular-weight DNA from soil microorganisms. Appl Environ Microbiol. 2008;74:3302–5.
[168] Leis B, Angelov A, Mientus M, et al. Identification of novel esterase-active enzymes from hot environments by use of the host bacterium *Thermus thermophilus*. Front Microbiol. 2015;6:275.
[169] Vester JK, Glaring MA, Stougaard P. Improved cultivation and metagenomics as new tools for bioprospecting in cold environments. Extremophiles. 2015;19:17–29.
[170] Liebl W, Angelov A, Juergensen J, et al. Alternative hosts for functional (meta)genome analysis. Appl Microbiol Biotechnol. 2014;98:8099–109.
[171] Ferrer M, Martinez-Martinez M, Bargiela R, Streit W, Golyshina O, Golyshin P. Estimating the success of enzyme bioprospecting through metagenomics: Current status and future trends. Microb Biotechnol. 2016;9:22–34.
[172] Tchigvintsev A, Tran H, Popovic A, et al. The environment shapes microbial enzymes: five cold-active and salt-resistant carboxylesterases from marine metagenomes. Appl Microbiol Biotechnol. 2015;99;2165–78.
[173] Ferrer M, Beloqui A, Vieites J, Guazzaroni M, Berger I, Aharoni A. Interplay of metagenomics and in vitro compartmentalization. Microb Biotechnol. 2009;2:31–9.
[174] Maruthamuthu M, Jiménez D, Stevens P, Van Elsas J. A multi-substrate approach for functional metagenomics-based screening for (hemi)cellulases in two wheat straw-degrading microbial consortia unveils novel thermoalkaliphilic enzymes. BMC Genomics. 2016;17:86.
[175] Smart et al. 2017.
[176] Buchholz PCF, Vogel C, Reusch W, et al. BioCatNet: A database system for the integration of enzyme sequences and biocatalytic experiments. ChemBioChem. 2016;17:2093–8.
[177] Krone M, Frieß F, Scharnowski K. et al. Molecular Surface Maps. IEEE Trans Visual Comput Graphics. 2017;23:701–10.
[178] Zeil C, Widmann M, Fademrecht S, Vogel C, Pleiss J. Network analysis of sequence-function relationships and exploration of sequence space of TEM β-lactamases. Antimicrob Agents Chemotherapy. 2016;60:2709–17.

[179] Verma D, Satyanarayana T. An improved protocol for DNA extraction from alkaline soil and sediment samples for constructing metagenomic libraries. Appl Biochem Biotechnol. 2011;165(2):454–64.

[180] Elcia M. S. Brito, Hilda A. Piñón-Castillo, Rémy Guyoneaud, César A. Caretta, J. Félix Gutiérrez-Corona, Robert Duran, Georgina E. Reyna-López, G. Virginia Nevárez-Moorillón, Anne Fahy, Marisol Goñi-Urriza. Bacterial biodiversity from anthropogenic extreme environments: a hyper-alkaline and hyper-saline industrial residue contaminated by chromium and iron. Applied Microbiology and Biotechnology 2013, 97 369–378.

[181] Vincent Scola, Jean-Baptiste Ramond, Aline Frossard, Olivier Zablocki, Evelien M Adriaenssens, Riegardt M Johnson, Mary Seely, Don A Cowan. Namib desert soil microbial community diversity, assembly, and function along a natural xeric gradient. Microbial Ecology 2018; 75 193–203.

[182] Yun Fang, Meiying Xu, Xingjuan Chen, Guoping Sun, Jun Guo, Weimin Wu, Xueduan Liu. Modified pretreatment method for total microbial DNA extraction from contaminated river sediment. Frontiers of Environmental Science & Engineering 2015; Volume 9, Issue 3, pp 444–452.

[183] Mitchell and Takacs-Vasbach. A comparison of methods for total community DNA preservation and extraction from various thermal environments. J Ind Microbiol Biotechnol. 2008 35(10):1139–47.

[184] Morono Y, Terada T, Hoshino T, Inagaki F. Hot-alkaline DNA extraction method for deep-subseafloor archaeal communities. Appl. Environ. Microbiol. 2014;80(6):1985–94.

[185] Jiang W, Liang P, Wang B, Fang J, Lang J, Tian G, Jiang J, Zhu TF. Optimized DNA extraction and metagenomic sequencing of airborne microbial communities. Nature protocols. 2015;10(5):768.

[186] Purdy KJ, Embley TM, Takii S, Nedwell DB. Rapid extraction of DNA and rRNA from sediments by a novel hydroxyapatite spin-column method. Appl Environ Microbiol. 1996;62:3905–3907.

[187] Valverde A, Tuffin M, Cowan DA. Biogeography of bacterial communities in hot springs: a focus on the actinobacteria. Extremophiles. 2012;16(4):669–79.

[188] Barton HA, Taylor NM, Lubbers BR, Pemberton AC. DNA extraction from low-biomass carbonate rock: an improved method with reduced contamination and the low-biomass contaminant database. Journal of Microbiological Methods. 2006;66(1):21–31.

[189] Ramond JB, Berthe T, Lafite R, Deloffre J, Ouddane B, Petit F. Relationships between hydrosedimentary processes and occurrence of mercury-resistant bacteria (merA) in estuary mudflats (Seine, France). Marine pollution bulletin. 2008;56(6):1168–76.

[190] Juniper SK, Cambon MA, Lesongeur F, Barbier G. Extraction and purification of DNA from organic rich subsurface sediments (ODP Leg 169S). Marine Geology. 2001;174(1–4):241–7.

Julien Tremblay

6 Practical overview of bioinformatics data mining in environmental genomics

Abstract: Over the past several years, the throughput of modern sequencing instruments has dramatically increased. This rapid increase in data production has put massive pressure on existing computing and storage infrastructure. High-performance computing (HPC) material is expensive and challenging to maintain. Many biotech businesses and governmental departments are willing to incorporate nucleic acid sequencing technology into their R&D pipelines, but can be discouraged and overwhelmed by the high costs of computing infrastructure and data analysis. This chapter will address good practices, methodology, and key concepts in the analysis of environmental genomics sequencing data of gene amplicons, shotgun metagenomics, and metatranscriptomics. A particular emphasis will be put on practical aspects of sequencing data processing.

6.1 Introduction

Early sequencing technologies of the 1970s enabled the sequencing of bacteriophages, viruses, mitochondria, and chloroplasts [1–5]. The first bacterial genome was later sequenced (Haemophilus Influenzae) in 1995 [6]. Around that time, the Sanger sequencing technology gained popularity and eventually achieved remarkable throughput for its time, allowing the simultaneous sequencing of hundreds of samples [7]. This technology enabled the completion of humanity's first major sequencing projects, with a human genome draft completed by 2001, followed by drafts of model organisms [8–10]. Roche's 454 technology was released in 2005 and is often considered as the first real high-throughput sequencing (HTS) technology that has been made available to scientists. Shortly after, the Illumina/Solexa corporation introduced molecular clustering technology sequencers. The adoption of these sequencing platforms by the scientific community coincided with a dramatic plunge in sequencing cost, which drove the democratization of nucleic acids sequencing technology. Sequencing cost steadily continued to drop during the last decade to reach a value of $US 0.1 per raw Mega-base (i.e. in 2017). This ever-increasing accessibility to sequencing technology combined to barcode indexing, which allows for the simultaneous sequencing of large number of samples [11, 12], resulted in an increased integration of DNA and RNA sequencing data in biology research projects. This in turn naturally resulted in generation of large datasets. Today, a typical sequencing project can generate up to terabytes of data, which renders its proper analysis challenging. In this regard, the appropriate and efficient analysis of this type of data requires large

https://doi.org/10.1515/9783110525786-006

computing infrastructure that is commonly known as a compute cluster. As biology is becoming more and more a data-intensive field of research, there is a need to develop standardized and systematic approaches to properly analyze modern HTS data.

Metagenomics is defined as the direct genetic analysis of genomes contained in an environmental sample [13] without the need for cultivation in the laboratory [14–16]. Investigation of extreme environments such as permafrost, acid mine drainage, deep subsurface, and extreme Antarctic generally makes use of three types of HTS products: (1) 16S/ITS rRNA marker gene amplicons, (2) shotgun metagenomics, and (3) shotgun metatranscriptomics. The former is a technique in which targeted fragments of the 16S or ITS rRNA gene are amplified using universal primers [17] followed by sequencing of the amplified material. The resulting data can then be processed using a plethora of established methods [14, 18] and generate results describing microbial communities (i.e. who's there and how abundant are they?). The 16S and ITS rRNA genes are highly conserved among archaea/bacteria prokaryotes and fungi, respectively, and are markers of choice to assess their phylogeny [19–21] and other community metrics [22] from an environmental sample. It is a well-established technique that has proven effective to survey the microbial populations in a sample without cultivation biases and efforts. It plays a central role in large ongoing microbial community studies such as the National Institutes of Health-funded Human Microbiome project [23, 24], and the Earth Microbiome project [25]. Due to its high cost and complexity, shotgun metagenomics has not reached the ubiquitousness of short rRNA gene amplicons. In contrast to rRNA gene amplicon sequencing, shotgun sequencing repertories every DNA molecule in a sample, not just one single gene (i.e. 16S rRNA gene or ITS region). Therefore, not only does this give information on phylogeny, but it also provides insights on all gene functions found in a given sample (i.e. who's there and who's doing what?). Shotgun metatranscriptomics consists of sequencing RNA transcripts directly from an environmental sample and can give insights on the intensity at which gene expression occurs in an environmental sample (i.e. who's there, how abundant are they, what are they doing and at what intensity they are doing it?). Shotgun metatranscriptomics usually involves a step where the highly abundant "contaminant" rRNA molecules that constitute the vast majority of total RNA extracts have to be removed prior to sequencing. This itself represents a significant challenge because high amounts of RNA molecules are required to construct sound sequencing libraries, to compensate for material loss during the rRNA molecules depletion step. This issue is absent in DNA-based sequencing approaches (shotgun metagenomics and rRNA gene amplicons) where commercial kits are available to construct high-quality sequencing library requiring from as low as one nanogram of starting material (purified DNA). This particular aspect of starting material input is important when planning RNA high-throughput sequencing projects on extreme environments as the amount of nucleic acids yielded from these types of environmental samples can be problematically low. In consequence, while interesting from a scientific point of view, shotgun metatranscriptomics can be challenging to perform, both from a logistic and financial perspective, because of the necessary rRNA depletion step.

6.2 Data types

At the moment of writing, the Illumina corporation is dominating the nucleic acids sequencing market with its short reads sequencing technology [26]. Data coming out of its HiSeq series line of instruments comes in the form of single-ended or paired reads of either 100, 150, 200, or 250 bp, while the MiSeq benchtop sequencer officially supports reads having to up to 300 bp in length. The complexity of modern genomics analyses is in part related to the inherent nature of these short reads: starting from short reads, one has to reach conclusions on a scientific question. Depending on the sequencing system, a typical sequencing lane of HiSeq can yield to up to 325 million sequence fragments. Here, it is important to clarify that a fragment actually represents two reads if the sequencing configuration is in paired-end mode. In that case, this would correspond to 650 million paired-end reads (2 × 325 million reads). In terms of raw reads output, Illumina's systems do not really have any competitors except perhaps from Oxford Nanopore's recent PromethION system, which, according to the company, can generate to up to a theoretical maximum of 11 Tera-bases of sequences in 48 hours (https://nanoporetech.com/products#comparison). Processing short reads is very computationally demanding. In a typical bioinformatics pipeline, there is often a step in which short reads are to be clustered or assembled into longer sequences or contigs. This step can require enormous amounts of RAM (Random Access Memory) [27] depending on the data complexity. Short reads can also be limiting in data interpretation when looking at subtle genetic processes such as RNA alternative splicing and accurately assembling contigs with GC-rich regions. The Pacific Biosciences corporation has worked toward filling that void and excelled into providing a long-reads sequencing technology (PacBio RSII) [28, 29] that can be applied to the field of metagenomics [30]. In contrast to Illumina's short reads (between 150 and 300 bp), long reads can range between 1 and 50 kb. The sequencing throughput of PacBio systems is orders of magnitude lower than what is obtained on an Illumina HiSeq system. Although small bacterial genomes can be assembled with high accuracy using PacBio data only [31], wisely using both short and long reads can optimize downstream analyses and is an approach that is increasingly gaining traction [32]. The reader is invited to consult the following references for in-depth reviews of high-throughput sequencing technologies [33–36].

6.3 Experimental design

The field of microbial ecology has historically lagged behind in terms of statistical robustness due to inadequate experimental designs [37, 38] and we cannot put enough emphasis on the fact that experimental design is paramount to achieve statistical significance and validate scientific hypotheses [39]. From the transcriptomics ("RNA-seq") field of study, an absolute minimum of biological triplicates [40] should be sequenced in order to achieve some level of statistical significance in gene differential abundance testing. However, other studies recommend that biological replicates should range between 6 and 12 [41, 42]. While not identical, transcriptomics and metagenomics sequencing data are comparable in many regards [43, 44], and assumptions about transcriptomics data behavior can be applied to metagenomics. Having an adequate number of replicates is vital to ensure that results found in downstream analyses are statistically sound and have real implication for the experimental planning phase of a project. For instance, the investigation of 10 treatments, including a control condition, means that 30 samples (10 treatments × 3 replicates) will have to be processed in the project pipeline. This can have a serious consequence on a restricted budget. Should budget limits be exceeded, dropping an experimental condition instead of lowering the number of replicates for one or multiple treatment conditions should be favored. A careful experimental planning can lead to interesting findings even when looking at a few experimental conditions only. On the other hand, having lots of underpowered samples cannot lead to statistical significance, which means there are no ways of assessing if observed results are outliers or within the limits of acceptable statistical variance. In the context of extreme environments, getting enough biological replicates for a given experimental condition can be challenging because of harsh sampling conditions, nature of sampled environment, availability of raw material, etc., and should be taken into account meticulously during project planning phase. Samples coming from extreme environments are highly valuable because of the costly logistics involved in their harvest: sampling trips in remote locations, specialized equipment, complex sampling methodologies, etc. It is consequently a good practice to harvest more biological replicates than needed as it is common to observe a failed DNA extraction or amplification for a given sample. Should a PCR or DNA extraction fail, having enough spare material in hand will prevent unfortunate situations where an experimental design becomes irrelevant due to lack of biological replicates and therefore avoid waste of resources.

16S/ITS rRNA gene amplicon surveys are relatively cheap to perform and can give extremely useful insights on a biological system. It should in all logic be the first step to achieve in an environmental genomics project. Then, based on the results obtained from these amplicons data, one can decide if it is worth it to perform additional amplicon, shotgun metagenomic, or even shotgun metatranscriptomic sequencing. Estimating the required reads coverage for 16S rRNA gene amplicon surveys is challenging. Some studies report the number of reads per sample to be between 1,000 and 20,000 [45–47]. Typical short amplicon profiling experiments should target around 20,000 fragments or 40,000 paired-end reads per sample to get enough depth to saturate or at least cover the most abundant microbes present in a sample. Reads depth per sample should be increased if the goal of the experiment is to find rare operational taxonomic units (OTUs), but in general, one should favor uniform rather than high depth. With the output of modern sequencing systems, such a sequencing depth is easily attainable [48, 49]. Targeted sequencing depth is directly correlated with the complexity of microbial communities found in the sampled environment: the more complex communities are, the more sequencing is required. Compared to matrices hosting complex communities like soils or wetlands, extreme environments are generally hosts to low-complexity microbial assemblages.

Shotgun metagenomic sequencing of DNA from an environmental sample is orders of magnitude more expensive than amplicon sequencing and has layers of additional challenges compared to shotgun sequencing of a single genome/organism. In eukaryotes genetics, reads are often mapped against a well annotated and optimized genome reference (i.e. human, mouse, chimp genomes). This means that we know exactly the location of each gene on each chromosome of that organism. This way, abundance profiles of each gene, that is the number of reads that mapped on a given gene of the reference genome, are obtained. Because we know beforehand how long the reference genome is and how many genes it contains, we can accurately estimate the number of reads to be sequenced in order to achieve enough coverage to reach statistical significance in downstream analyses (reviewed in [50]). For instance, targeted coverage for a higher organism can easily reach to up to 100×. In other words, each base of the genome is, on average, covered by 100 bases from 100 different reads. In contrast, when performing amplicon or shotgun metagenomic sequencing experiments, we do not know which microbes are present in our samples. Because this crucial information is missing, this complicates coverage estimates as a typical environmental sample can contain millions of microbes, each with their own chromosomes and plasmids, themselves harboring numerous genes. By their inherent nature, the complexity of samples coming from extreme environments, such as contaminated soil, acid mine drainage [51–54], deep subsurface [48, 55–57], and extreme Antarctic [58, 59], are usually of low complexity and reads coverage should therefore be less of a problem. On the other hand, other environments such as wetlands and soil rhizosphere can also be considered extreme because of their remarkably high complexity [27, 60, 61]. In such environmental matrices, reads coverage can be problematic and reaching complete or near-complete sequencing saturation is often intractable and it is therefore impractical to cover all organisms with an adequate sequencing coverage. Deciding which sequencing depth to target per sample is often a blind shot. In order to reconcile budget constraints and scientific questions, our group usually favors sequencing between 2 to 6 Giga-bases per sample for typical projects investigating at least 24 samples. This gives enough overall sequencing depth to obtain a decent metagenome reference co-assembly and enough depth to perform quantitative tests on most abundant genes and microbes. Should the number of sample be low, for instance 6 to 12, sequencing output per sample should be substantially increased in order to obtain a good co-assembly reference for most abundant microorganisms.

6.4 Short rRNA gene amplicons

Because of its low cost, robust databases and established bioinformatics procedures, short amplicon sequencing of a rRNA gene fragment (not to confound with shotgun metagenomics) is usually the first step in a project aiming at characterizing microbial populations from a given environment. This type of experiment enables to determine which microbes are present and how relatively abundant they are. Based on the relevance of these microbial communities characterization, an informed decision regarding the need to perform shotgun metagenomic and/or metatranscriptomic sequencing can be made. A wealth of literature is available for description of procedures for short rRNA marker genes bioinformatics processing [47, 62–64]. Here, I will describe the procedure I use in production in our research unit at the National Research Council of Canada (Fig. 6.1).

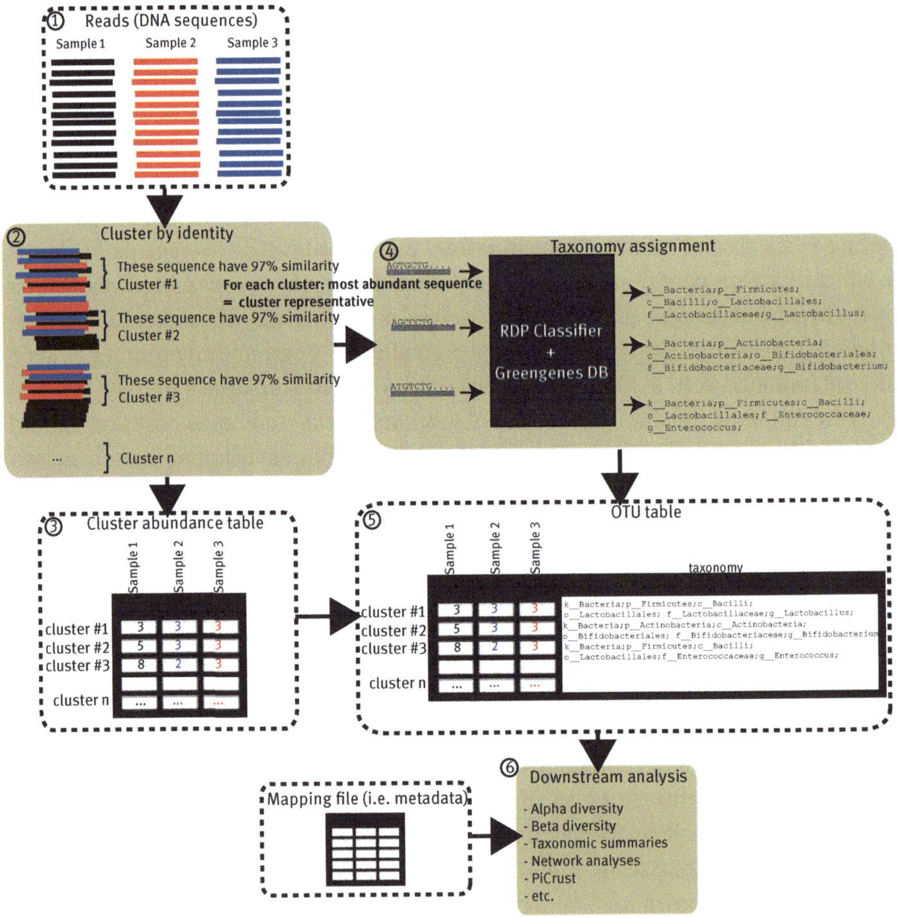

Fig. 6.1: Overview of a typical rRNA gene amplicon bioinformatics pipeline. Raw reads are filtered for quality (1). QC-passed reads are clustered by identity (2). Resulting clusters are reported in an abundance table showing the number of reads in each cluster for each sample (3). Clusters are assigned a taxonomic lineage using a reference database (4) to generate an OTU table (5). Using metadata, various analyses are performed from the OTU table (6).

First of all, which hypervariable region and, consequently, which primer pair is to be used to generate amplicons needs to be decided. Numerous primer pairs targeting various hypervariable regions, along with their corresponding biases, have been described in the literature [65–70]. Typically, studies focusing on environments such as soil and plant rhizosphere/endosphere relies on the Earth microbiome project V4/V4–V5 primers [71–73]. In contrast, human microbiome-related studies lean more towards the use of the V1–V3, V3–V5, and V4–V5 regions [23, 24]. Biases generated during laboratory work (e.g. DNA extraction, PCR, and library preparation methods) are increasingly being documented [74–76] and should be considered as well.

Once samples have been collected, DNA extraction, amplification of the chosen hypervariable region, and library preparation are performed on samples that are to be investigated. The MiSeq (Illumina) benchtop sequencer is suitable for most short amplicon marker gene studies, and multiplexing up to 384 samples on a MiSeq lane can yield enough reads per sample for most types of environments. Typically, soil and rhizosphere samples are highly complex, and completely covering their diversity is difficult to achieve. A real example of a MiSeq sequencing run is described in Tab 6.1. In that particular case, an average of 95,000 fragments per sample was obtained by multiplexing 170 samples on one sequencing lane. As stated above, this should be more than enough to effectively cover most abundant OTUs and adequately perform downstream analyses across samples [77, 78].

Once a sequencing run is completed, fastq files are generated. These files are available in many configurations, but typically, one fastq file for forward reads and one fastq file for reverse reads are generated. Fastq files can further be demultiplexed using the barcode index information associated with each read. This way, we end up with one fastq file per read orientation per sample.

With fastq files in hands, data processing can begin.

1. Reads are first scanned for contaminants (e.g. Illumina, 454, or PacBio adapter sequences) and PhiX reads using the DUK software (unpublished, http://duk.sourceforge.net). Usually, a small proportion of reads are contaminants, and accordingly, 0–25% are PhiX reads.
2. Removal of unpaired reads. From step 1, paired-end reads may be disrupted. This means that one of the read in a pair might be lost due to the screening in step 1. All of these unpaired reads are discarded. This is usually a fairly small proportion of all reads. This step is not performed if reads are single ended (MiSeq single ended, PacBio, IonTorrent, or 454).
3. All reads are then trimmed to a fixed length that is variable depending on the quality of the sequencing run and amplicon length. It is a good practice to trim reads to a fixed length before downstream analyses, especially if reads are single-end (i.e. 454, IonTorrent, or PacBio data types). If reads are paired end, 3' trimming should be done in a way that enough bases are left on each read

pairs to allow assembly using forward (reads 1) and reverse (reads 2) common overlapping parts.

4. If paired-end reads: Reads are assembled (overlapping paired assembly) with the FLASH software [79].
5. Primer sequences may or may not be removed from the assembled/single-end reads. Primer sequences should be removed when possible/applicable as the primer annealing regions of amplified DNA may be overrepresented in sequencing errors (personal observations).
6. The trimmed assembled/single-end reads from step #4–5 are filtered for quality. All reads having an average quality score lower than 33 or more than 1 N (undefined base) and 5 nucleotides below quality 15 are discarded. The remaining reads will be referred to as filtered reads from now on.
7. Filtered reads are then clustered with our in-house clustering workflow. Briefly, reads are clustered at 100% identity and then clustered/denoised at 99% identity (dnaclust [79, 80]). Clusters having abundance lower than 3 are discarded. Remaining clusters are then scanned for chimeras with UCHIME *de novo* and UCHIME reference [81] and clustered at 97% (dnaclust) to form the final clusters/OTUs.
8. OTUs are assigned a taxonomy. Briefly, OTUs are classified with the Ribosomal Database Project (RDP) classifier [82] using an in-house training set containing the complete Greengenes database [83] supplemented with eukaryotic sequences from the Silva database and a customized set of mitochondria and chloroplasts 16S sequences. The ITS2 database consists of the UNITE ITS database (ITS1–ITS2). The RDP classifier gives a score (0 to 1) for each taxonomic depth of each OTU. Each taxonomic depth having a score ≥0.5 is kept to reconstruct the final lineage. Clusters/OTUs are also blasted against the most recent NCBI nt database for complementary information.
9. Using taxonomic lineages obtained from step 8 combined to cluster/OTU abundance from step 7, a raw OTU table is generated. From that raw OTU table, an OTU table containing both bacterial and archeal organisms is generated. From this latter OTU table, a normalized OTU table (edgeR [43, 84]) is generated [43]. If data consist of ITS amplicons, the same procedures are applied, but the raw OTU table is filtered to keep fungal organisms only.
10. A summary of reads counted throughout the different steps of the pipeline is generated. This is useful to get a global outlook on the sequencing run: how many reads were sequenced, how many reads filtered out after QC, how many clusters/OTUs were generated, etc.
11. From these classified OTUs, diversity metrics are obtained by aligning OTU sequences on a Greengenes core reference alignment [83] using the PyNAST aligner [85]. Alignments are filtered to keep only the hypervariable region part of the alignment.

Tab. 6.1: Example for a typical MiSeq run output on a 2 × 250 bp configuration.

Output type	Characteristic
Output (Giga-bases)	7.32 Giga-bases
Output (millions of fragments)	16,319,951 fragments
Output (millions of reads)	2 × 16,319,951 = 32,639,902 reads
Average output per sample with 170 indexes/barcodes	16,319,951 fragments / 170 = 95,999 fragments per sample
Sequencing output	runxxx_sampleA_R1.fastq.gz (contains forward reads of sample A) runxxx_sampleA_R2.fastq.gz (contains reverse reads of sample A) runxxx_sampleB_R1.fastq.gz (contains forward reads of sample B) runxxx_sampleB_R2.fastq.gz (contains reverse reads of sample B) ...

12. A phylogenetic tree is then built from that alignment (from step 11) with FastTree [85, 86]. Alpha (observed species) and beta (weighted, unweighted UniFrac, and Bray Curtis distances) diversity metrics and taxonomic summaries are then computed using the QIIME software suite [63, 85]. Along with the OTU tables, these last tables represent end results from the pipeline and can then be used to generate various types of plots and statistics computation (Fig. 6.1).

One of the most critical steps in a rRNA amplicon pipeline is arguably the OTU generation step, also known as OTU-picking. OTU generation methodology is a complex topic addressed elsewhere [18, 87–89]. There are mainly five databases for taxonomic assignment: NCBI rRNA, RDP, Greengenes, Silva, and the Open Tree of Life Taxonomy [90]. The RDP classifier is widely used for classifying reads or OTU sequences, but the RDP web portal/database is often confused with the RDP classifier software [82], which is an implementation of a Bayesian classifier that can be used with any customized training sets (e.g. databases). From personal observations, the RDP classifier performs well with exhaustive training sets. In my bioinformatics procedures, I typically use the RDP classifier in combination with a customized database consisting of the latest Greengenes or Silva [91] (respectively v13_5 and 128 at the moment of writing) 16S rRNA genes databases. Chloroplast and mitochondrial sequences and eukaryotes 18S rRNA sequences from the Silva database can also be added to the reference 16S rRNA gene database. Integrating these contaminant 16S/18S rRNA sequences into a training set/database is critical for assigning proper taxonomic lineage to OTUs coming from samples where there can be many of these non-desired rRNA gene sequences. For instance, samples coming from a plant rhizosphere compartment will typically have many plastid 16S sequences in their sequencing libraries. These plastid reads can form OTUs, and if no proper plastid sequences are included in the reference training set used for classifying OTUs, they can be assigned the closely related *Cyanobacteria* taxon. The amount of these undesired amplicons can also be minimized upstream of data analysis during amplification by using peptide nucleic acid PCR clamps developed to suppress plant host plastid and mitochondrial 16S contaminants [92]. Some universal 16S rRNA gene primers can also amplify 18S eukaryotes rRNA genes (V6–V8 primers in [69, 93]), which also need to be flagged and excluded from OTU tables.

One limitation of short 16S rRNA amplicon sequencing is the low resolution in the obtained taxonomic lineages. For instance, DNA fragments amplified with primers targeting the V4 hypervariable region will generate amplicons of 290 bp, which becomes ~252 bp after removal of primer annealing sites. Accordingly, OTU representative sequences will also have an average length of ~252 bp. A fair proportion of V4 OTUs can be confidently assigned to up to the genus level but often can only go as far as the class, order, or family levels. One way to circumvent this is to generate amplicons with primers englobing the full 16S rRNA gene (in contrast to amplifying only a short hypervariable region) and get them sequenced on a Pacific Biosciences system that generates long reads, which can in turn be used to generate and classify OTUs to up to the species level [94–96] because of their high resolution (~1,500 bp vs. ~252 bp). This sequencing technology effectively allows the high-quality sequencing of the whole 16S rRNA amplified gene. However, even by having the full rRNA gene length, database limitations can prevent full classification of PacBio amplicon data. Because of the considerably lower sequencing depth and higher cost of doing amplicon sequencing on a PacBio system, this technology is seldom used for amplicon sequencing.

A great wealth of information can be pulled out of amplicon data. Classic data analysis stream consists of using the OTU table to generate taxonomic summaries, alpha- and beta-diversity metrics, OTU network analyses and functional profile predictions [18, 97, 98]. Using appropriate statistical procedures, differentially abundant OTUs between two treatment conditions can also be determined [44].

Using amplicon sequencing data from a phytoremediation project, we show a typical stream of analyses (Fig. 6.2) starting from an OTU table. OTU distribution can first be observed with a heatmap (Fig. 6.2a). A heatmap can quickly inform on the global patterns of OTU abundance across samples. For the current example, it is clear from the heatmap that samples form two distinct clusters based on the contamination variable (contaminated vs. non-contaminated): many OTUs that are highly abundant in contaminated soil are scarce in non-contaminated soil and vice versa. Alpha-diversity metrics inform on the complexity of microbial communities in a given experimental condition (e.g. how many different OTUs are present in a sample and how they are distributed). In this example (Fig. 6.2b), contaminated samples had lower diversity than uncontaminated samples. Beta-diversity metrics inform on the proximity of samples from one another- in other words, how different is sample *x* from sample *y*. Weighted UniFrac distance [99] and Bray-Curtis

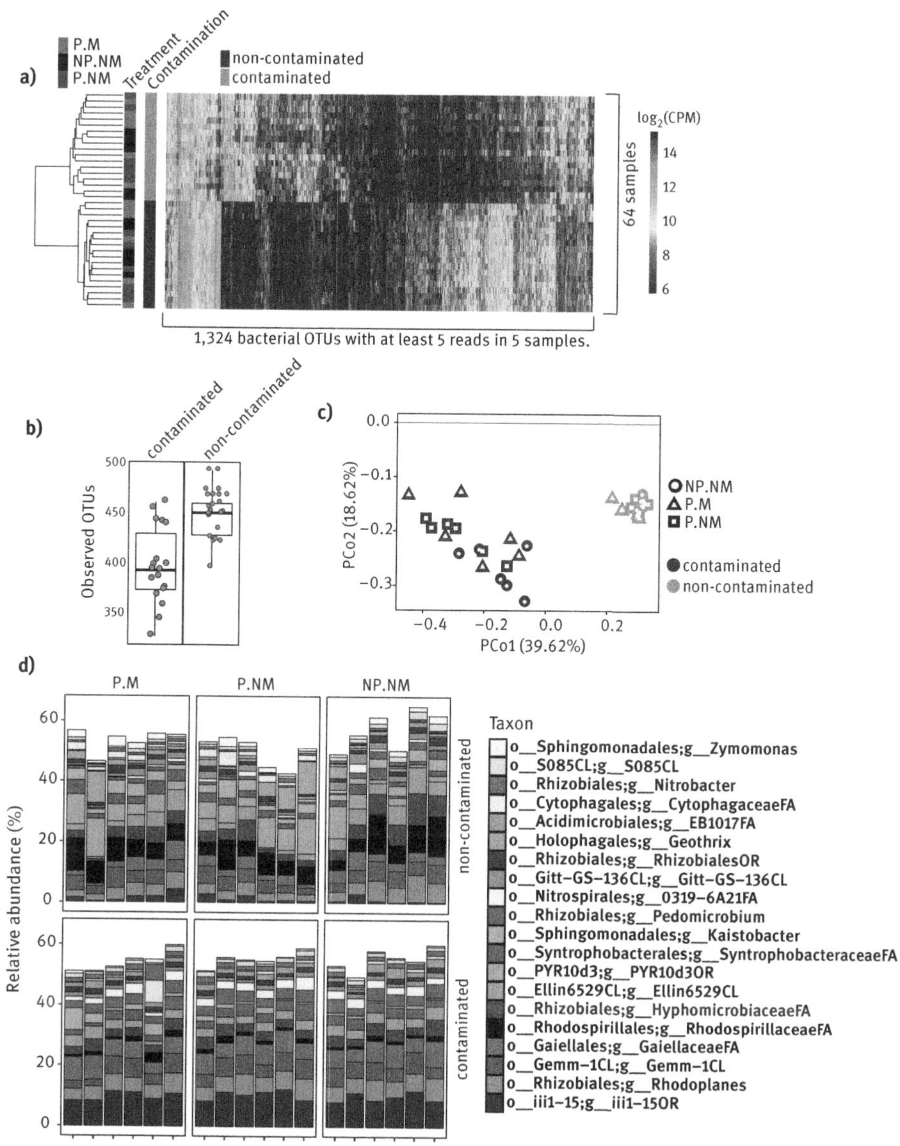

Fig. 6.2: Typical microbial community analyses done with rRNA gene amplicon data. (a) Heatmap containing all OTUs having at least five reads in five samples (out of a total of 60 samples). Each column represents an OTU and each row represent a sample, so that each cell indicates the abundance of an OTU in a given sample. (b) Alpha-diversity boxplots with the observed OTU index, where each dot represents the estimated diversity index of a sample. (c) Principal coordinates (PCoA) of weighted UniFrac distances between all samples. (d) Taxonomic profiles binned by contamination and treatment variables. Note that the abundance of each taxon is an aggregation of taxonomic lineages assigned to each OTU. o__ = taxon at the order level; g__ = taxon at the genus level; P.M = planted with mycorrhizal fungi; NP.NM = nonplanted without mycorrhizal fungi; P:NM = planted without mycorrhizal fungi; CPM = count per million.

dissimilarity [100] indexes, which are widely used metrics to assess beta-diversity for microbial communities, are computed from the OTU table. Weighted UniFrac incorporates OTU abundance and phylogenetic distance as input parameters, while Bray-Curtis relies solely on OTUs abundance. The resulting distance or dissimilarity matrix is transformed using a principal coordinates transformation so that each sample can be visualized on two- or three-dimensional axes. This process of transforming a distance matrix into coordinates is termed principal coordinates analysis (PCoA). The PCoA in Fig. 6.2c shows that samples cluster primarily by their contamination state (contaminated vs. uncontaminated), which is in accordance with the clustering patterns observed in the heatmap of Fig. 6.2a. Plotting taxonomic profiles enables the identification of the taxonomic group(s) affected by a given experimental condition (Fig. 6.2d). In our example, differences in taxa profiles can be observed between the P.M, P.NM, and NP.NM groups inside each contaminated and uncontaminated groups. In this case, however, taxonomic profiles are more importantly affected between the contaminated and uncontaminated experimental variable. For instance, the *iii-1-15* and *Rhodospirillaceae* groups are orders of magnitude more abundant in contaminated soil, while *SO85* is abundant in uncontaminated soil but virtually absent in contaminated soil (Fig. 6.2d). Here, the shown taxonomic profiles include only the 20 most abundant taxa. These 20 taxa constitute roughly 50% of all taxa and offer only a partial picture of all the shifts in taxa abundance profiles. While useful, showing barplots to describe microbial communities is quite restrictive, as the number of colors are limiting to efficiently distinguish microbial taxa from one another. An alternative way to show differences in taxonomic profiles is to first identify all OTUs that are significantly differentially abundant between two conditions [44] and then perform taxonomic summarization of these identified OTUs. Many other types of analyses can be done to further put forward differences in experimental treatments and characterize microbial communities. The reader is invited to consult existing literature on the topic for further information [13, 14, 18, 22, 23].

6.5 Metagenomic shotgun sequencing

As stated earlier, shotgun metagenomic data is, on many levels, more complex to analyze and interpret than short rRNA gene amplicon data. A typical workflow of shotgun metagenomic data analysis is illustrated in Fig. 6.3.

The following is a summarized description of our internal shotgun metagenomics analysis pipeline [101].

1. Quality control. Sequencing adapters are removed from each read. Reads are head-cropped at 13 bp (i.e. the first 12–13 bases of each read are prone to sequencing errors and are overrepresented in either A, C, G, or Ts [Trimmomatic]). Reads are also filtered using a sliding window approach: starting from their 3' region, each read is scanned with a four bases wide sliding window, cutting the read when the average quality drops below Phredscore 15. Finally, after these filtering

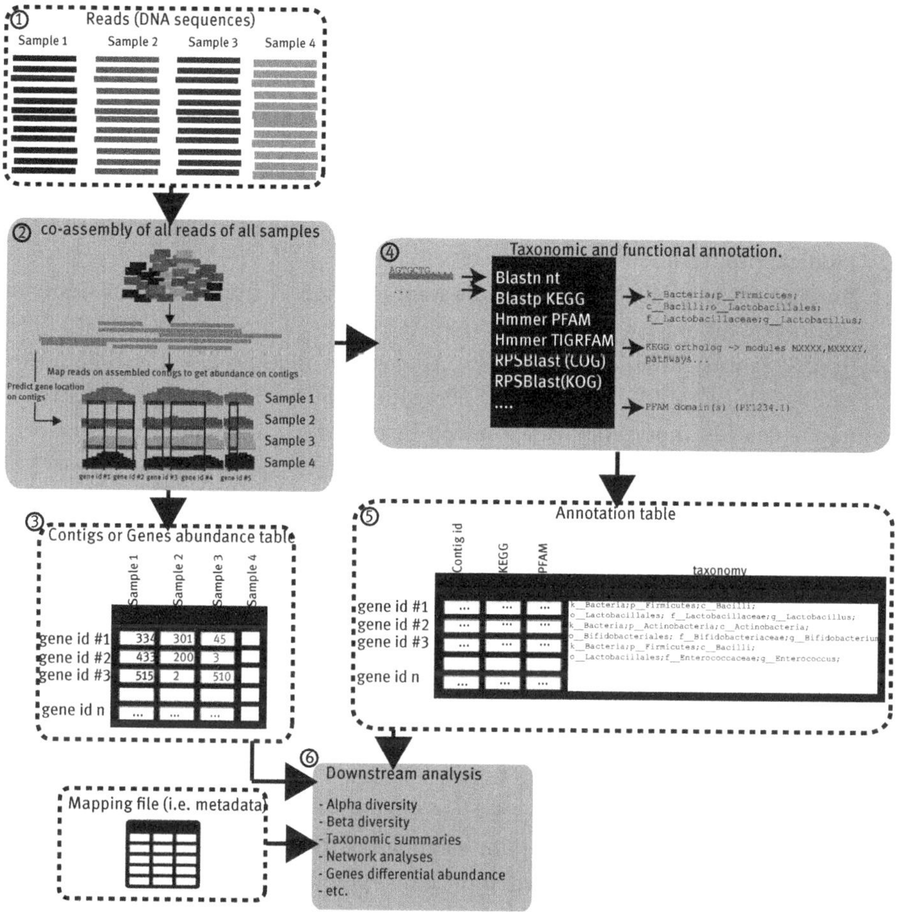

Fig. 6.3: Overview of a typical shotgun metagenomics bioinformatics pipeline. Reads are first filtered for quality (1) and co-assembled into contigs (2). QC-passed reads (the ones used for assembly) are mapped against contigs to obtain contig abundance profiles (2). Genes coordinates on each contig is determined using a gene prediction model (2) so that abundance profiles of genes can be computed from contig abundance profiles. Annotation is done by comparing each gene sequence against various databases to get functional and taxonomic information associated to each gene (4). With the abundance (3), annotation (5), and metadata tables in hand, downstream analyses (6) can be done to focus on scientific questions.

procedures, reads having lower than 50 bases in length are discarded. This step is done using Trimmomatic [102]. Appropriate QC is critical to obtain an optimal assembly and reads mapping in downstream steps.

2. Contaminant removal. Reads are scanned for contaminants (e.g. Illumina, 454, or PacBio adapter sequences) and PhiX reads using the DUK software (unpublished, http://duk.sourceforge.net). Usually, a small proportion of reads are contaminants, and accordingly, 0–25% are PhiX reads.

3. Reads assembly into contigs. Each QC-passed read of each sample is co-assembled using the Megahit [103] or Ray [104, 105] software. If I have access to a large memory compute node (1 terabyte RAM or more), Megahit will be favored. If that's not the case, Ray will be used on enough compute nodes to meet the amount of RAM required for the assembly.
4. Gene prediction on obtained contigs. This is done by calling genes (ORF, start codon, stop codon, etc.) on each assembled contig using MetageneMark [106] or Prodigal [107, 108].
5. Annotation of predicted genes. This step is inspired from the Joint Genome Institute's annotation guidelines [109]: (1) RPSBLAST against COG DB, (2) RPS-BLAST against KOG DB, (3) HMMSCAN against PFAM-A DB, (4) HMMSCAN against TIGRFAM DB, and (5) BLASTP against KEGG DB. For taxonomy assignment, contigs (not genes) are blasted against NCBI's nt database, and the best hit having an e-value ≤ 1e-02 and alignment length of at least 90 bases is considered.
6. Mapping (BWA mem; unpublished, http://bio-bwa.sourceforge.net) of QC-passed reads against contig sequences. QC-passed reads are then mapped against contigs to assess quality of metagenome assembly and to obtain abundance profiles of all samples.
7. Metagenome binning. Contigs from (3) and abundance profiles obtained in (5) are used as input for the MetaBAT [110, 111] binning software. Obtained metagenome bins are analyzed for quality (CheckM [111]) and can be further parsed using in-house scripts [101].
8. Statistics. According to the supplied experimental design, differential DNA abundance is computed (edgeR [84]) for each design. Genes having a false discovery rate (FDR) value ≤ 0.05 and log(Fold-Change) (logFC) ≥ 1.5 are considered to be differentially abundant. Cutoffs can be parameterized for specific needs. Note that many packages exist to test for differentially abundant genes from metagenomic datasets [112].
9. Taxonomic summary. Taxonomic summaries are computed (QIIME [63, 85]) for taxonomy based on both contigs and bins.

With shotgun metagenomic data, similar results as the ones reported in Fig. 6.2 can be obtained. Heatmaps of bin, contig, or gene abundance can be generated. Taxonomic summaries using contig or bin abundance can be profiled. Alpha- and beta-diversity can be computed based on either contig, gene, or bin abundance. Of course, results do not have the same fundamental meaning as OTU-based results, but they are conceptually similar. The real power of metagenomic data lies in assessing the functional profiles of genes across experimental conditions. Various databases exist for this purpose. Here, we'll focus on using the Kyoto Encyclopedia of Genes and Genomes (KEGG) database [113–115]. The basal unit in the KEGG ecosystem is the KEGG orthology (i.e. KO number K*12345*). Each ortholog is associated with three types of molecular interactions: KEGG pathway maps, BRITE hierarchies, and KEGG modules. KO numbers are assigned to genes in reference genomes. By comparing gene sequences from a metagenomic sequencing dataset against KEGG reference genes, it is possible to determine which KO identifier is associated with the obtained hit. From there, the KO identifier can be used to infer which modules or pathways is a particular gene part of. In Fig. 6.4, I show an example of a KEGG module profile overrepresentation analysis using shotgun metagenomic sequencing data from 12 samples from a phytoremediation project. In this example, we have all differentially abundant genes (FDR $\leq$ 0.05 and logFC $\geq$ 1.5) between contaminated and non-contaminated conditions that match the KEGG modules part of the ABC-2 type and other transport systems group. Fig. 6.4a shows that genes belonging to the module M00254 were highly abundant in the contaminated compared to non-contaminated soil samples. Similar trends are observed for modules M00250, M00259, M00255, and M00256. The combination of taxonomic and functional information for each gene gives powerful insights on what microorganisms are actually involved in selected metabolic processes (i.e. who's doing what?). Fig. 6.4b shows that *Pseudomonas* is overrepresented in the K01990 ortholog in contaminated samples, while K01992 genes are associated with *Azoarcus*. Many genes in which K01990 and K01992 homologies were found were not assigned a taxonomic lineage due to the absence of significant homology against public databases.

6.6 Metatranscriptomic shotgun sequencing

Shotgun metatranscriptomic sequencing consists of sequencing mRNA instead of DNA. This tells us about the expression and not only the presence or abundance of a gene. While shotgun metagenomics enables answering who's doing what?, metatranscriptomics goes a step further and aims at answering who's doing what and at what intensity. Analyzing metatranscriptomic data is, in many ways, similar to metagenomic data analysis. Starting from raw reads, there are mainly two routes that can be taken: (1) *de novo* assembly of raw reads, followed by annotation and gene abundance estimation – essentially what is described in Fig. 6.3, but using sequenced RNA instead of DNA – or (2) map (*i.e.* align) RNA QC-passed reads against a previously assembled and annotated metagenome reference (generated by the shotgun metagenomics pipeline), ideally done using the same set of samples. Metatranscriptome data are usually less complex than metagenome data, because it only includes expressed transcripts of a biological system at the moment of RNA extraction. In contrast, metagenome data includes not only expressed genes but also all nucleic acid material from chromosomes, plasmid intergenic regions, etc. As a consequence, genome assembly software needs less memory to successfully assemble metatranscriptomic data because their implemented k-mer hash structures are far less voluminous.

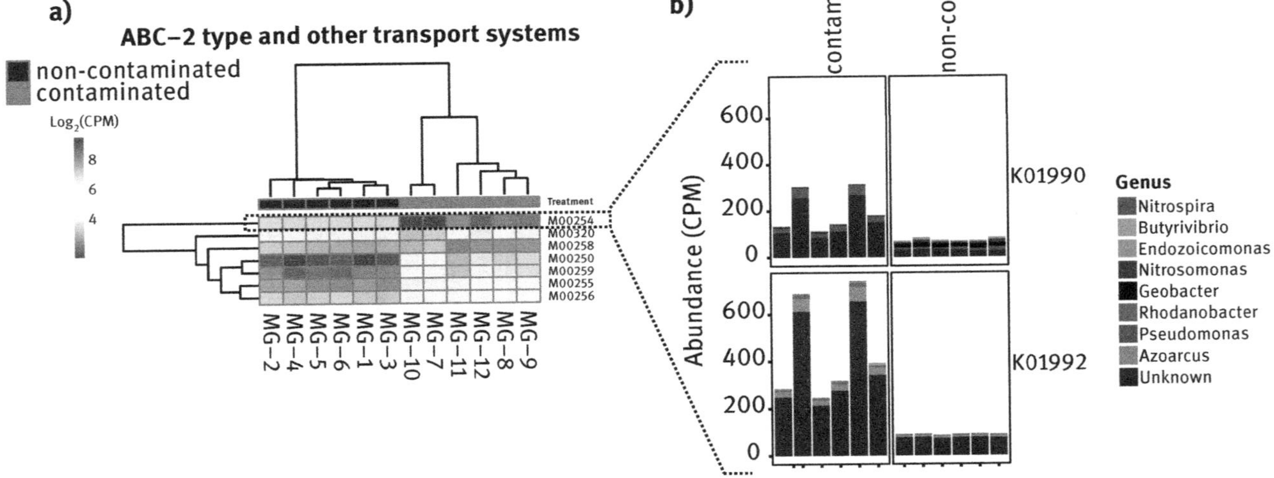

Fig 6.4: KEGG module enrichment analysis done with shotgun metagenomic sequencing data. (a) Heatmap with average clustering where columns correspond to samples and rows to KEGG modules. Each cell corresponds to the aggregated $\log_2$(CPM) of each differently abundant genes (assessed with edgeR) assigned with the corresponding KEGG module. (b) Detailed KEGG ortholog abundance profile for module M00254 for each taxon.

6.7 Additional notes

6.7.1 Challenges of annotation

One culprit of shotgun 'omics analyses is that lots of contigs and genes are often impossible to annotate because of the absence of homology against public databases. For instance, in complex environments like soil, it is common to have more than 60% of unannotated genes. This means that we may end up with sequences in our dataset that are probably real, but for which we have no idea to which organism they belong or what metabolic function they might be associated with. In many cases, genes are partially annotated: they might have high similarity hits against one particular database, but none against others. For instance, a given gene may very well show strong similarity to a known KEGG ortholog involved in *x* metabolic pathway or module but does not show enough similarity to any listed microbe to enable a confident taxonomic assignation.

6.7.2 Pipeline wrapper

Bioinformatics pipelines can be complex, with many steps that need to be executed in a specific order. In order to gain productivity, pipelines should ideally be executed on a compute cluster using a job scheduler (e.g. Torque, SLURM) supporting job dependencies. This way, a 1,000-job pipeline can be submitted all at once to the job scheduler so that each job can be available for execution only once their depending job have successfully completed. For example, in a shotgun metagenomic pipeline, the co-assembly job can enter the waiting queue only when the quality control jobs it depends on have all been successfully completed. Only then the co-assembly job will enter the queue and wait for its execution. Many workflow management systems or pipeline wrappers (software that generates scripts of job submissions) have been written and published (reviewed in [116]) and others are being developed in-house, like the McGill University and Genome Quebec Innovation Centre pipeline module (https://bitbucket.org/mugqic/mugqic_pipelines). A good pipeline wrapper module should generate jobs, manage their dependencies, and have a smart restart mechanism in case of job failure. In the context of a complex pipeline with thousands of jobs, a smart restart mechanism is indispensable to gain productivity and save time determining which job failed. For instance, in annotation procedures, the multi-fasta file holding gene sequences of the co-assembled contigs is usually split in smaller chunks so that instead of having a single big annotation job that compares 2 million gene entries in a single job, many smaller annotation jobs are submitted in parallel to the job scheduler. In the event that one of these smaller annotation jobs fails due to hardware instability or maintenance, manually finding which job exactly failed can be a tedious task. With a smart restart mechanism implementation, the pipeline wrapper should find, upon re-execution, which job actually failed to successfully complete and effectively rewrite them for resubmission. Pipeline wrappers are also critical in that they allow sequencing data to be systematically analyzed in reproducible ways and that each job that it generates is parameterizable. For instance, when analyzing quality controlled reads, one can realize that the head-crop length of 12 bases specified for Trimmomatic was too short. By increasing the head-crop value to 15 bases in a separate parameter file and rerunning the pipeline wrapper, jobs integrating this modified parameter will be regenerated and resubmitted to the job scheduler. Proceeding with a pipeline wrapper also leaves traces of parameters used in all jobs should the data and analyses be revisited in the future. Operating a pipeline wrapper requires sizable skills at the command line on a Linux-based system, which is not necessarily at everyone's reach. In this context, bioinformatics platforms integrating graphical user interfaces provide alternatives [117–119] to conduct analyses.

6.8 Future directions

Biology is steadily incorporating the character of a data science discipline. This is especially true in the high-throughput sequencing field of metagenomics, where a single sample contains millions of microbial genomes themselves harboring thousands of genes, which are composed of hundreds to thousands of nucleotides. These challenges are amplified by the fact that the data coming out of most modern sequencers are in the form of short reads that need significant computational processing before being brought to an intelligible format. Next-generation sequencers are already pushing towards generating longer reads, which should in theory soften the data processing inherent to short reads. Yet, the continuous drop in sequencing cost will probably drive the demand for more samples being sequenced, therefore increasing the already high dimensionality of environmental genomics data.

References

[1] Sanger F, Air GM, Barrell BG, Brown NL, Coulson AR, Fiddes CA, et al. Nucleotide sequence of bacteriophage phi X174 DNA. Nature. 1977;265:687–95.

[2] Sanger F, Coulson AR, Hong GF, Hill DF, Petersen GB. Nucleotide sequence of bacteriophage lambda DNA. J Mol Biol. 1982;162:729–73.

[3] Bankier AT, Beck S, Bohni R, et al. The DNA sequence of the human cytomegalovirus genome. DNA Seq. 1991;2:1–12.

[4] Oda K, Yamato K, Ohta E, et al. Gene organization deduced from the complete sequence of liverwort *Marchantia polymorpha* mitochondrial DNA. A primitive form of plant mitochondrial genome. J Mol Biol. 1992;223:1–7.

[5] Ohyama K, Fukuzawa H, Kohchi T, et al. Chloroplast gene organization deduced from complete sequence of liverwort *Marchantia polymorpha* chloroplast DNA. Nature. 1986;322:572–4.
[6] Fleischmann RD, Adams MD, White O, et al. Whole-genome random sequencing and assembly of *Haemophilus influenzae* Rd. Science. 1995;269:496–512.
[7] Ansorge WJ. Next-generation DNA sequencing techniques. N Biotechnol. 2009;25:195–203.
[8] Mouse Genome Sequencing Consortium, Waterston RH, Lindblad-Toh K, et al. Initial sequencing and comparative analysis of the mouse genome. Nature. 2002;420:520–62.
[9] Gibbs RA, Weinstock GM, Metzker ML, et al. Genome sequence of the Brown Norway rat yields insights into mammalian evolution. Nature. 2004;428:493–521.
[10] Analysis Consortium TCS, The Chimpanzee Sequencing and Analysis Consortium. Initial sequence of the chimpanzee genome and comparison with the human genome. Nature. 2005;437:69–87.
[11] Andersson AF, Lindberg M, Jakobsson H, Bäckhed F, Nyrén P, Engstrand L. Comparative analysis of human gut microbiota by barcoded pyrosequencing. PLoS One. 2008;3:e2836.
[12] Hamady M, Walker JJ, Harris JK, Gold NJ, Knight R. Error-correcting barcoded primers for pyrosequencing hundreds of samples in multiplex. Nat Methods. 2008;5:235–7.
[13] Thomas T, Gilbert J, Meyer F. Metagenomics – a guide from sampling to data analysis. Microb Inform Exp. 2012;2:3.
[14] Hiraoka S, Yang C-C, Iwasaki W. Metagenomics and bioinformatics in microbial ecology: current status and beyond. Microbes Environ. 2016;31:204–12.
[15] Pace NR. A molecular view of microbial diversity and the biosphere. Science. 1997;276:734–40.
[16] Rabus R, Venceslau SS, Wöhlbrand L, Voordouw G, Wall JD, Pereira IAC. A Post-genomic view of the ecophysiology, catabolism and biotechnological relevance of sulphate-reducing prokaryotes. Adv Microb Physiol. 2015;66:55–321.
[17] Tringe SG, Hugenholtz P. A renaissance for the pioneering 16S rRNA gene. Curr Opin Microbiol. 2008;11:442–6.
[18] Hugerth LW, Andersson AF. Analysing microbial community composition through amplicon sequencing: from sampling to hypothesis testing. Front Microbiol. 2017;8. doi:10.3389/fmicb.2017.01561.
[19] Weisburg WG, Barns SM, Pelletier DA, Lane DJ. 16S ribosomal DNA amplification for phylogenetic study. J Bacteriol. 1991;173:697–703.
[20] Olsen GJ, Lane DJ, Giovannoni SJ, Pace NR, Stahl DA. Microbial ecology and evolution: a ribosomal RNA approach. Annu Rev Microbiol. 1986;40:337–65.
[21] Ward DM, Weller R, Bateson MM. 16S rRNA sequences reveal numerous uncultured microorganisms in a natural community. Nature. 1990;345:63–5.
[22] Birtel J, Walser J-C, Pichon S, Bürgmann H, Matthews B. Estimating bacterial diversity for ecological studies: methods, metrics, and assumptions. PLoS One. 2015;10:e0125356.
[23] Human Microbiome Project Consortium. A framework for human microbiome research. Nature. 2012;486:215–21.
[24] Human Microbiome Project Consortium. Structure, function and diversity of the healthy human microbiome. Nature. 2012;486:207–14.
[25] Gilbert JA, Meyer F, Antonopoulos D, et al. Meeting report: the terabase metagenomics workshop and the vision of an Earth microbiome project. Stand Genomic Sci. 2010;3:243–8.
[26] Timmerman L. DNA sequencing market will exceed $20 billion, says Illumina CEO Jay Flatley [Internet]. 2015. Available at: http://www.forbes.com/sites/
[27] Georganas E, Buluç A, Chapman J, et al. HipMer: an extreme-scale de novo genome assembler. Proceeding SC '15 Proceedings of the International Conference for High Performance Computing, Networking, Storage and Analysis. Austin, TX, USA; IEEE; 15–20 Nov 2015.
[28] Eid J, Fehr A, Gray J, et al. Real-time DNA sequencing from single polymerase molecules. Science. 2009; Jan 2; 323(5910):133–8.

[29] Levene MJ, Korlach J, Turner SW, Foquet M, Craighead HG, Webb WW. Zero-mode waveguides for single-molecule analysis at high concentrations. Science. 2003;299:682–6.
[30] Bowman B, Kim M, Cho Y-J, Korlach J. Long-read, single molecule, real-time (SMRT) DNA sequencing for metagenomic applications. In: Metagenomics for microbiology. 2015. p. 25–38.
[31] Chin C-S, Alexander DH, Marks P, et al. Nonhybrid, finished microbial genome assemblies from long-read SMRT sequencing data. Nat Methods. 2013;10:563–9.
[32] Liao Y-C, Lin S-H, Lin H-H. Completing bacterial genome assemblies: strategy and performance comparisons. Sci Rep. 2015;5:8747.
[33] van Dijk EL, Auger H, Jaszczyszyn Y, Thermes C. Ten years of next-generation sequencing technology. Trends Genet. 2014;30:418–26.
[34] Kuo PC. Mechanisms of Next-Generation Sequencing (NGS). In: Next-generation sequencing and sequence data analysis. Bentham Science Publishers Ltd; Sharjah, U.A.E. Author; 2015. p. 25–37.
[35] Heather JM, Chain B. The sequence of sequencers: the history of sequencing DNA. Genomics. 2016;107:1–8.
[36] Goodwin S, McPherson JD, McCombie WR. Coming of age: ten years of next-generation sequencing technologies. Nat Rev Genet. 2016;17:333–51.
[37] Prosser JI. Replicate or lie. Environ Microbiol. 2010;12:1806–10.
[38] Webster R. Replicate and randomize, or lie. Environ Microbiol. 2016;19:25–8.
[39] Knight R, Jansson J, Field D, et al. Unlocking the potential of metagenomics through replicated experimental design. Nat Biotechnol. 2012;30:513–20.
[40] Soneson C, Delorenzi M. A comparison of methods for differential expression analysis of RNA-seq data. BMC Bioinformatics. 2013;14:91.
[41] Schurch NJ, Schofield P, Gierliński M, et al. How many biological replicates are needed in an RNA-seq experiment and which differential expression tool should you use? RNA. 2016;22:839–51.
[42] Burden CJ, Qureshi SE, Wilson SR. Error estimates for the analysis of differential expression from RNA-seq count data. PeerJ. 2014;2:e576.
[43] McMurdie PJ, Holmes S. Waste not, want not: why rarefying microbiome data is inadmissible. PLoS Comput Biol. 2014;10:e1003531.
[44] Weiss S, Xu ZZ, Peddada S, et al. Normalization and microbial differential abundance strategies depend upon data characteristics. Microbiome. 2017;5:27.
[45] Martínez I, Stegen JC, Maldonado-Gómez MX, et al. The gut microbiota of rural Papua New Guineans: composition, diversity patterns, and ecological processes. Cell Rep. 2015;11:527–38.
[46] Obregon-Tito AJ, Tito RY, Metcalf J, et al. Subsistence strategies in traditional societies distinguish gut microbiomes. Nat Commun. 2015;6:6505.
[47] Lundberg DS, Lebeis SL, Paredes SH, et al. Defining the core Arabidopsis thaliana root microbiome. Nature. 2012;488:86–90.
[48] Yergeau E, Michel C, Tremblay J, et al. Metagenomic survey of the taxonomic and functional microbial communities of seawater and sea ice from the Canadian Arctic. Sci Rep. 2017;7:42242.
[49] Piché-Choquette S, Tremblay J, Tringe SG, Constant P. H2-saturation of high affinity H2-oxidizing bacteria alters the ecological niche of soil microorganisms unevenly among taxonomic groups. PeerJ. 2016;4:e1782.
[50] Sims D, Sudbery I, Ilott NE, Heger A, Ponting CP. Sequencing depth and coverage: key considerations in genomic analyses. Nat Rev Genet. 2014;15:121–32.
[51] Baker BJ, Banfield JF. Microbial communities in acid mine drainage. FEMS Microbiol Ecol. 2003;44:139–52.
[52] Kuang J-L, Huang L-N, Chen L-X, et al. Contemporary environmental variation determines microbial diversity patterns in acid mine drainage. ISME J. 2013;7:1038–50.

[53] Tyson GW, Chapman J, Hugenholtz P, et al. Community structure and metabolism through reconstruction of microbial genomes from the environment. Nature. 2004;428:37–43.
[54] Baker BJ, Tyson GW, Webb RI, et al. Lineages of acidophilic archaea revealed by community genomic analysis. Science. 2006;314:1933–5.
[55] Yergeau E, Greer CW. Metagenomic analysis of polar ecosystems. In: Polar microbiology: life in a deep freeze. Robert V Miller, Lyle G Whyte, Editors. ASM Press, Washington DC. p. 156–65.
[56] Collins RE, Rocap G, Deming JW. Persistence of bacterial and archaeal communities in sea ice through an Arctic winter. Environ Microbiol. 2010;12:1828–41.
[57] Bowman JS, Rasmussen S, Blom N, Deming JW, Rysgaard S, Sicheritz-Ponten T. Microbial community structure of Arctic multiyear sea ice and surface seawater by 454 sequencing of the 16S RNA gene. ISME J. 2012;6:11–20.
[58] Lopatina A, Medvedeva S, Shmakov S, Logacheva MD, Krylenkov V, Severinov K. Metagenomic analysis of bacterial communities of Antarctic surface snow. Front Microbiol. 2016;7:398.
[59] Lopatina A, Krylenkov V, Severinov K. Activity and bacterial diversity of snow around Russian Antarctic stations. Res Microbiol. 2013;164:949–58.
[60] He S, Malfatti SA, McFarland JW, et al. Patterns in wetland microbial community composition and functional gene repertoire associated with methane emissions. MBio. 2015;6:e00066-15.
[61] Howe AC, Jansson JK, Malfatti SA, Tringe SG, Tiedje JM, Brown CT. Tackling soil diversity with the assembly of large, complex metagenomes. Proc Natl Acad Sci U S A. 2014;111:4904–9.
[62] Lax S, Smith DP, Hampton-Marcell J, Owens SM, Handley KM, Scott NM, et al. Longitudinal analysis of microbial interaction between humans and the indoor environment. Science. 2014;345:1048–52.
[63] Kuczynski J, Stombaugh J, Walters WA, González A, Caporaso JG, Knight R. Using QIIME to analyze 16S rRNA gene sequences from microbial communities. Curr Protoc Bioinformatics. 2011;Chapter 10:Unit 10.7.
[64] Edgar RC. UPARSE: highly accurate OTU sequences from microbial amplicon reads. Nat Methods. 2013;10:996–8.
[65] Klindworth A, Pruesse E, Schweer T, Peplies J, Quast C, Horn M, et al. Evaluation of general 16S ribosomal RNA gene PCR primers for classical and next-generation sequencing-based diversity studies. Nucleic Acids Res. 2013;41:e1.
[66] Lee CK, Herbold CW, Polson SW, Wommack KE, Williamson SJ, McDonald IR, et al. Groundtruthing next-gen sequencing for microbial ecology – biases and errors in community structure estimates from PCR amplicon pyrosequencing. PLoS One. 2012;7:e44224.
[67] Pinto AJ, Raskin L. PCR biases distort bacterial and archaeal community structure in pyrosequencing datasets. PLoS One. 2012;7:e43093.
[68] He Y, Zhou B-J, Deng G-H, Jiang X-T, Zhang H, Zhou H-W. Comparison of microbial diversity determined with the same variable tag sequence extracted from two different PCR amplicons. BMC Microbiol. 2013;13:208.
[69] Tremblay J, Singh K, Fern A, et al. Primer and platform effects on 16S rRNA tag sequencing. Front Microbiol. 2015;6:771.
[70] Ibarbalz FM, Pérez MV, Figuerola ELM, Erijman L. The bias associated with amplicon sequencing does not affect the quantitative assessment of bacterial community dynamics. PLoS One. 2014;9:e99722.
[71] Caporaso JG, Lauber CL, Walters WA, et al. Ultra-high-throughput microbial community analysis on the Illumina HiSeq and MiSeq platforms. ISME J. 2012;6:1621–4.
[72] Takahashi S, Tomita J, Nishioka K, Hisada T, Nishijima M. Development of a Prokaryotic universal primer for simultaneous analysis of bacteria and Archaea using next-generation sequencing. PLoS One. 2014;9:e105592.
[73] Parada AE, Needham DM, Fuhrman JA. Every base matters: assessing small subunit rRNA primers for marine microbiomes with mock communities, time series and global field samples. Environ Microbiol. 2016;18:1403–14.

[74] Kennedy K, Hall MW, Lynch MDJ, Moreno-Hagelsieb G, Neufeld JD. Evaluating bias of illumina-based bacterial 16S rRNA gene profiles. Appl Environ Microbiol. 2014;80:5717–22.
[75] Brooks JP, Edwards DJ, Harwich MD Jr, et al. The truth about metagenomics: quantifying and counteracting bias in 16S rRNA studies. BMC Microbiol. 2015;15:66.
[76] Marotz C, Amir A, Humphrey G, Gaffney J, Gogul G, Knight R. DNA extraction for streamlined metagenomics of diverse environmental samples. Biotechniques. 2017;62:290–3.
[77] Hamady M, Knight R. Microbial community profiling for human microbiome projects: tools, techniques, and challenges. Genome Res. 2009;19:1141–52.
[78] Falony G, Joossens M, Vieira-Silva S, et al. Population-level analysis of gut microbiome variation. Science. 2016;352:560–4.
[79] Magoč T, Salzberg SL. FLASH: fast length adjustment of short reads to improve genome assemblies. Bioinformatics. 2011;27:2957–63.
[80] Ghodsi M, Liu B, Pop M. DNACLUST: accurate and efficient clustering of phylogenetic marker genes. BMC Bioinformatics. 2011;12:271.
[81] Edgar RC, Haas BJ, Clemente JC, Quince C, Knight R. UCHIME improves sensitivity and speed of chimera detection. Bioinformatics. 2011;27:2194–200.
[82] Wang Q, Garrity GM, Tiedje JM, Cole JR. Naive Bayesian classifier for rapid assignment of rRNA sequences into the new bacterial taxonomy. Appl Environ Microbiol. 2007;73:5261–7.
[83] DeSantis TZ, Hugenholtz P, Larsen N, et al. Greengenes, a chimera-checked 16S rRNA gene database and workbench compatible with ARB. Appl Environ Microbiol. 2006;72:5069–72.
[84] Robinson MD, McCarthy DJ, Smyth GK. edgeR: a bioconductor package for differential expression analysis of digital gene expression data. Bioinformatics. 2010;26:139–40.
[85] Caporaso JG, Kuczynski J, Stombaugh J, et al. QIIME allows analysis of high-throughput community sequencing data. Nat Methods. 2010;7:335–6.
[86] Price MN, Dehal PS, Arkin AP. FastTree 2 – approximately maximum-likelihood trees for large alignments. PLoS One. 2010;5:e9490.
[87] Westcott SL, Schloss PD. De novo clustering methods outperform reference-based methods for assigning 16S rRNA gene sequences to operational taxonomic units. PeerJ. 2015;3:e1487.
[88] Schloss PD. Application of a database-independent approach to assess the quality of operational taxonomic unit picking methods. mSystems. 2016;1. doi:10.1128/mSystems.00027-16.
[89] Kopylova E, Navas-Molina JA, Mercier C, Xu ZZ, Mahé F, He Y, et al. Open-source sequence clustering methods improve the state of the art. mSystems. 2016;1. doi:10.1128/mSystems.00003-15.
[90] Balvočiūtė M, Huson DH. SILVA, RDP, Greengenes, NCBI and OTT – how do these taxonomies compare? BMC Genomics. 2017;18:114.
[91] Quast C, Pruesse E, Yilmaz P, et al. The SILVA ribosomal RNA gene database project: improved data processing and web-based tools. Nucleic Acids Res. 2013;41:D590–6.
[92] Lundberg DS, Yourstone S, Mieczkowski P, Jones CD, Dangl JL. Practical innovations for high-throughput amplicon sequencing. Nat Methods. 2013;10:999–1002.
[93] Kunin V, Engelbrektson A, Ochman H, Hugenholtz P. Wrinkles in the rare biosphere: pyrosequencing errors can lead to artificial inflation of diversity estimates. Environ Microbiol. 2010;12:118–23.
[94] Singer E, Bushnell B, Coleman-Derr D, et al. High-resolution phylogenetic microbial community profiling. ISME J. 2016;10:2020–32.
[95] Wagner J, Coupland P, Browne HP, Lawley TD, Francis SC, Parkhill J. Evaluation of PacBio sequencing for full-length bacterial 16S rRNA gene classification. BMC Microbiol. 2016;16:274.
[96] Franzén O, Hu J, Bao X, Itzkowitz SH, Peter I, Bashir A. Improved OTU-picking using long-read 16S rRNA gene amplicon sequencing and generic hierarchical clustering. Microbiome. 2015;3:43.

[97] Langille MGI, Zaneveld J, Caporaso JG, et al. Predictive functional profiling of microbial communities using 16S rRNA marker gene sequences. Nat Biotechnol. 2013;31:814–21.
[98] Aßhauer KP, Wemheuer B, Daniel R, Meinicke P. Tax4Fun: predicting functional profiles from metagenomic 16S rRNA data. Bioinformatics. 2015;31:2882–4.
[99] Lozupone C, Knight R. UniFrac: a new phylogenetic method for comparing microbial communities. Appl Environ Microbiol. 2005;71:8228–35.
[100] Bray JR, Roger Bray J, Curtis JT. An ordination of the upland forest communities of southern Wisconsin. Ecol Monogr. 1957;27:325–49.
[101] Tremblay J, Yergeau E, Fortin N, et al. Chemical dispersants enhance the activity of oil- and gas condensate-degrading marine bacteria. ISME J. 2017; doi:10.1038/ismej.2017.129.
[102] Bolger AM, Lohse M, Usadel B. Trimmomatic: a flexible trimmer for Illumina sequence data. Bioinformatics. 2014;30:2114–20.
[103] Li D, Liu C M, Luo R, Sadakane K, Lam T-W. MEGAHIT: an ultra-fast single-node solution for large and complex metagenomics assembly via succinct de Bruijn graph. Bioinformatics. 2015;31:1674–6.
[104] Boisvert S, Laviolette F, Corbeil J. Ray: simultaneous assembly of reads from a mix of high-throughput sequencing technologies. J Comput Biol. 2010;17:1519–33.
[105] Boisvert S, Raymond F, Godzaridis E, Laviolette F, Corbeil J. Ray Meta: scalable de novo metagenome assembly and profiling. Genome Biol. 2012;13:R122.
[106] Zhu W, Lomsadze A, Borodovsky M. Ab initio gene identification in metagenomic sequences. Nucleic Acids Res. 2010;38:e132–e132.
[107] Hyatt D, Chen G-L, Locascio PF, Land ML, Larimer FW, Hauser LJ. Prodigal: prokaryotic gene recognition and translation initiation site identification. BMC Bioinformatics. 2010;11:119.
[108] Hyatt D, LoCascio PF, Hauser LJ, Uberbacher EC. Gene and translation initiation site prediction in metagenomic sequences. Bioinformatics. 2012;28:2223–30.
[109] Huntemann M, Ivanova NN, Mavromatis K, et al. The standard operating procedure of the DOE-JGI Metagenome Annotation Pipeline (MAP v.4). Stand Genomic Sci. 2016;11:17.
[110] Kang DD, Froula J, Egan R, Wang Z. MetaBAT, an efficient tool for accurately reconstructing single genomes from complex microbial communities. PeerJ. 2015;3:e1165.
[111] Parks DH, Imelfort M, Skennerton CT, Hugenholtz P, Tyson GW. CheckM: assessing the quality of microbial genomes recovered from isolates, single cells, and metagenomes. Genome Res. 2015;25:1043–55.
[112] Jonsson V, Österlund T, Nerman O, Kristiansson E. Statistical evaluation of methods for identification of differentially abundant genes in comparative metagenomics. BMC Genomics. 2016;17:78.
[113] Kanehisa M, Furumichi M, Tanabe M, Sato Y, Morishima K. KEGG: new perspectives on genomes, pathways, diseases and drugs. Nucleic Acids Res. 2017;45:D353–61.
[114] Kanehisa M. KEGG: Kyoto Encyclopedia of Genes and Genomes. Nucleic Acids Res. 2000;28:27–30.
[115] Kanehisa M, Sato Y, Kawashima M, Furumichi M, Tanabe M. KEGG as a reference resource for gene and protein annotation. Nucleic Acids Res. 2016;44:D457–62.
[116] Leipzig J. A review of bioinformatic pipeline frameworks. Brief Bioinform. 2017;18:530–6.
[117] Li P-E, Lo C-C, Anderson JJ, et al. Enabling the democratization of the genomics revolution with a fully integrated web-based bioinformatics platform. Nucleic Acids Res. 2017;45:67–80.
[118] Shringarpure SS, Carroll A, De La Vega FM, Bustamante CD. Inexpensive and highly reproducible cloud-based variant calling of 2,535 human genomes. PLoS One. 2015;10:e0129277.
[119] Afgan E, Baker D, van den Beek M, blankenberg d, bouvier d, čech m, et al. The Galaxy platform for accessible, reproducible and collaborative biomedical analyses: 2016 update. Nucleic Acids Res. 2016;44:W3–10.

Itumeleng Moroenyane and Étienne Yergeau

7 Techniques and approaches to quantify microbial diversity in extreme environments

7.1 Summary

Soil diversity is a fundamental component of ecosystem functioning, but standardizing approaches to quantify it has been a challenge. Current studies rely on the use of compositional data to quantify the abundances of dominant microbial taxa across sites, and few still rely on traditional taxonomic approaches alone for soil microfauna. However, there is no consensus on what are the "best practices" for quantifying the soil biodiversity. High-throughput sequencing technologies have enabled the enumeration and comparison of previously unknown soil taxa from complex soil ecosystems by using relatively conserved regions of the genome (16S rRNA gene-bacterial and archaea, 18S rRNA gene-metazoa and fungi, and ITS region [for fungi]).

7.2 Introduction

There exists an intrinsic link between definitions of what a species is and how biodiversity can be measured; however, there is yet to be a unifying species concept for microbes, and this makes it harder to account for the immense microbial diversity [1]. These limitations are further compounded by lack of intuitive understanding that diversity is not an ecological processes but really a by-product of ecological processes [2]. These challenges extend beyond just the species concept, but also in characterizing the functional diversity of soil microbial communities [3]. The use and classification of ecological stable strategies for soil microbes have become a useful tool for ecologists. Fierer et al. [4] proposed a new system that classifies microbial taxa by their carbon utilization and niche occupancy – that is, microbes that use and survive in carbon-rich environments are classified as copiotrophs, while those found in carbon-limited environments are oligotrophs.

In soil ecosystems, this system becomes problematic as environmental filtering (influences of environment on the vertical and horizontal distribution of microbes) is often correlated with variations and estimates of soil diversity [5, 6]. And it seems in part that the biogeographical patterns of soil microbes are delimited not only by carbon utilization but also ecological processes that influence macro-organisms [7]. Furthermore, the use of trait-based approaches to quantify microbial diversity becomes concerning when the frequency of genetic material transfer (mobile genetic elements, horizontal gene transfer, and occurrence of environmental DNA being incorporated by living cells) in soils is still unknown [8]. Additionally, the physical

https://doi.org/10.1515/9783110525786-007

properties of soils change over spatial and temporal scales, and understanding how microbial communities are separated from one another or within their pore microhabitats is crucial to unlocking the mysteries of soil microbial diversity [9].

The field of microbial ecology has moved beyond just being descriptive (describing microbial communities and identifying which edaphic variables delimit their biogeography) [10, 11] and has adopted a more mechanistic approach (identifying evolutionary processes that are important in assembling communities) [12, 13]. These studies have elucidated the mechanism by which niche-based (environmental selection) and neutral processes (ecological drift, dispersal limitation, and random speciation events) influence microbial community biogeography and dynamics. Similar approaches have shown that bacterial communities from recently exposed soils following a glacial retreat are highly influenced by pH and not successional age, and pH seems to mediate the balance between niche-based and neutral processes [14]. Additionally, there has been an increasing number of studies that have investigated the aboveground-belowground dynamics, specifically how the soil microbiome plays an integral part in the functioning of terrestrial ecosystems and its role in improving soil fertility [15], soil formation [16], and decomposition [17]. An extensive review by Wardle et al. [18] first highlighted the intrinsic link that exists between aboveground and belowground; they proposed that belowground food webs directly and indirectly influence plant community succession, which in turn modulates the belowground microbial community through rhizodeposition and plant litter. The influences of plant community on soil microbial community have been shown to extend beyond the rhizosphere and extend beyond the canopy of the plant [19]. These trends were observed in the high Arctic tundra, where undisturbed soils were shown to possess higher levels of bacterial diversity than Boreal forests [20, 21]. Equally, in the high Antarctic, microbiota enumerated from soil or soil-like substrate have been shown to be important components of ecosystem functioning [22], to possess incredible diversity [23], and to occupy highly specialized niche spaces [24]. However, studies have highlighted that Rapoport's rule, i.e. species abundance and ranges are higher at lower latitudes and lower at higher latitudes, has been shown to hold for microbes in Antarctica [25]. However, it has been shown that it is more than the influence of latitude that predicts this diversity but strong environmental filtering, especially pH [26, 27].

These studies and more have highlighted the immense diversity of soil biota in these seemingly inhospitable environments. However, the majority of these studies have quantified the microbial diversity using different sampling scales, techniques, diversity indices, and sampling approaches, i.e. culture-dependent and culture-independent methods. Thus, there are contrasting data on what really delimits the distribution of microbes in these environments. Despite these challenges, there are few studies that have tried to elucidate the global distribution and diversity of microbes [28–30]. Moreover, microbial diversity is often reported out of context and incorrectly interpreted [2, 31]; much of this arises from microbial ecology "borrowing" ecological concepts of diversity from macro (traditional) ecology. For instance, the interpretation of microbial diversity indices and their ecological interpretation are still vastly unknown. An emerging question in the field is – what happens when basic principles such as species concept begin to breakdown and what usefulness is there in describing species richness? Even the units of diversity – alpha (α; i.e. the number of species in a local environment), beta (β; i.e. is the total turnover in species between two local environments), and gamma (γ; i.e. the total diversity of landscape and incorporates both α and β) – have, in recent years, come under scrutiny [32, 33]. Therefore, it is our belief that a comprehensive statistical approach that consolidates current and proposed methods can be applied to make the most of next-generation sequencing (NGS) microbial community data.

Historically, much of the work on description of microbial communities in extreme environmental habitats was restricted by access to study sites and sample preparation. Remote sampling, increased access to sites, and advancements in sequencing technologies and bioinformatics approaches have led to a drastic increase in published studies from these extreme environments (Fig. 7.1).

This increase is attributable, in part, to the replacement of culture-dependent cultivation methods by culture-independent NGS technologies and increased access to these extreme sites [34]. Furthermore, the relative cost of sequencing has declined significantly while depth of coverage continues to increase exponentially, and these factors, compounded with drastic decreases in error rates, have made it possible for researchers to survey microorganisms with larger genomes [35]. Microbial diversity estimations are often generated from amplicon datasets, i.e. the amplification and sequencing of a conserved region on the microbial genome, and the majority of studies use 16S rRNA marker gene for bacterial and archaeal components and 18S rRNA marker gene for most eukaryotes along with the fungal ITS (internal transcribed region). These amplicon studies account for the large proportion of published studies on microbial diversity. Recently, it has become possible to sequence larger genomes and do broad-scale metagenome sequencing of soil microorganisms. It seems that NGS technologies are a double-edged sword: on one side, they highlight the immense diversity of soil biota by circumventing traditional approaches, and on the other, the large output of data generated from NGS platforms has brought along new set of challenges [36, 37]. For example, NGS produces a lot of raw data, and this imposes limitations on data mining and interpretation, especially with increased sample size. A consequence of this large size of data produced highlights the storage limitation/constraints in computing storage and need for new curated databases/depositories to collate all this data. One contentious issue with NGS dataset is its compositional nature – that is, sequence output (total number of reads) in its underived state is inherently uninformative, and significance can be derived only from abundance ratios of the different components [38, 39]. As a result of this intrinsic nature of NGS datasets, relative abundance is often used as an index to quantify microbial diversity. However, the use of relative abundance as an accurate index for diversity has been a controversial issue in macroecology [40–42] and recently in microbial ecology [2, 28]. An important notion to understand when interpreting relative abundance is that it does not convey any real relationships,

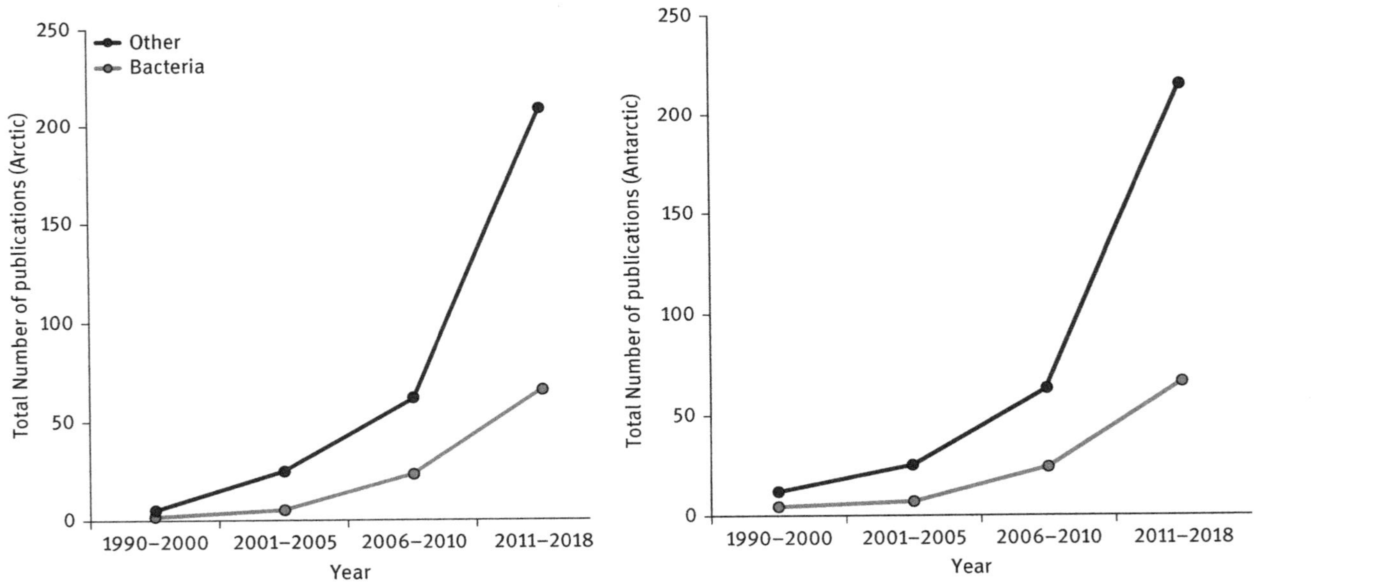

Fig. 7.1: Trends in the number of publications from the Arctic and Antarctic regions. The data represent annual number of publications on soil microbial diversity (bacterial and other, which includes nematodes, protist, fungi, and any microorganism). Searches were made on Web of Science, Scopus, PubMed, and Agricola databases on March 15, 2018. To search for article related to the Arctic we searched by topic = ((microbe* OR prokaryote* OR eukaryote* OR virus OR bacteria) AND diversity AND arctic AND (soil OR permafrost)), and for Antarctic =((microbe* OR prokaryote* OR eukaryote* OR virus OR bacteria) AND diversity AND antarctic AND (soil OR permafrost)). For articles that were solely focused on bacteria, we searched for articles that had the term "bacteria/bacterial" in the title.

i.e. the real abundance of microbes in soils is often incongruent to relative abundance. The change in absolute abundance has strong biological significance, but these changes can be treated as a simple artefact of sequencing; for example, a 20-fold change in the abundance of a dominant gene is likely to have a significant interpretation, whereas a 20-fold increase in expression of an extremely low abundant gene might just be an artefact of either the sequencing processes or bioinformatic pipeline. In many cases, the use of relative abundance in microbial studies has to be accompanied by quantification of absolute abundances (Real-Time Quantitative Polymerase chain reaction *i.e.* rt-qPCR, qPCR, and use of internal standards) and the relative abundance of each sample can be corrected using these standards. This approach has already been adopted and has been shown to greatly increase the value of compositional data in soils [43]. Additionally, the use of flow cytometry coupled with relative abundance has been shown to generate microbial abundance counts [43, 44]. New statistical approaches that have been proposed offer some resolution; these include, but are not limited to, new regression analysis of microbiome compositional data [45] and bioinformatics pipelines [38]. Both these methods resolve the fundamental compositional nature of NGS dataset by allowing the identification of taxa that is associated with a specific response even when data are summarized as compositional at different taxonomic levels. This is achieved in two ways: a regression analysis that considers these compositional data as covariates and linear models with constraints on the regression coefficients are incorporated. This approach has been demonstrated to detect microbial niche shifts [46] and disentangling complex gut microbial communities [47]. Despite all these new statistical approaches, it is the method of Smets et al. [48] of using internal genomic standards during sample preparations that shows potential. The introduction of internal standards, i.e. the addition of a known concentration of DNA from a taxon that is unlikely to be present in soils, e.g. *Alivibrio fischeri*, prior to DNA extraction of the soils sample improves the normalisation and quantification of DNA samples. This spike of foreign DNA allowed for the correct estimation of microbial biomass, and results were congruent with more traditional approaches, i.e. phospholipid fatty acid and substrate induced respiration. Thus far, this approach seems to be the most viable option for assessing the abundance and diversity of soil microorganisms.

In this book chapter, we will discuss the current approaches used to quantify microbial diversity in extreme environments and highlight new approaches that could be adopted.

7.3 Current approaches used to quantify soil microbial diversity in extreme environments

7.3.1 Microbial molecular genetic markers

Molecular genetic markers, such as the bacterial and archaeal 16S rRNA gene, are widely used to assess the richness, diversity, and function of soil microorganism

[49, 50]. The 16S rRNA genetic marker has become the hallmark of microbial research, and its attractiveness is partly due to the existence of extensive databases that are available for the correct identification of microbial taxa, e.g. Greengenes [51], EzTaxon-e [52], EzBioCloud [53], and many others. Similarly, there are markers that are widely used to characterize and assess the diversity of micro-eukaryotes in soil environments. Typically, 18S rRNA gene and CO1 are preferred genetic markers for most eukaryotes, while ITS is used for fungal identification. The ease and reliability of these markers have shaped our understanding of the complex nature of many environments [54].

In all cases, the target genetic marker is amplified using polymerase chain reaction (PCR), and the amplicons are sequenced for identification and characterization. The sequence data are then processed through the bioinformatic pipeline to remove chimera and check sequence quality with the goal of identifying unique sequences that are correctly matched/mapped to a known taxon. Generally, for prokaryotes, these unique sequences often are considered as operational taxonomic units (OTUs), and these are based on a clustering of similar sequences that meet a particular threshold (typically 97% sequence similarity), whereas for eukaryotes, molecular OTUs (MOTUs) are used at various thresholds (typically 99% sequence similarity). In both instances, the identification of a "species" is related to the molecular variations that exist in the marker gene. However, integrated operation taxonomic units (IOTUs) are gaining popularity. IOTUs differ significantly from MOTUs/OTUs in that they integrate at least one taxonomic characteristic of the taxa [55]. The use of sequence similarities as proxy for species in a microbial context has inherent limitations, such as different species tend to be clustered as a single OTU and ambiguity in "appropriate" species cutoffs. Recently, the use of amplicon sequence variants (analogous to 100% sequence similarity OTUs) as proxy for species is becoming the gold standard in microbial research, i.e. assigning each unique sequences to a taxon [56].

7.3.2 Measuring local species pools and turnover rates across sites

Alpha (α)-diversity

Alpha-diversity is fundamentally characterized by two intrinsic properties of biological communities, that is, richness and evenness. Of these properties, richness (measure of number of taxa found in the particular environment) is often used as an indicator of local species pool, while evenness (variations in the number of individuals per taxon in local species pool) is often recorded in studies that investigate microbial functions and interactions. Typically, microbial diversity is quantified using indices that combine both properties (richness and evenness), such as Shannon (a semibalanced index that has been shown to be biased toward dominant taxa in sample), Simpson (a semibalanced index that has been shown to be biased toward rare taxa in sample), and Chao (an index that estimates the total number of species in a sample given the abundance of individuals therein) [57]. These indices provide some insight on what the community dynamics are by combining diversity properties (richness and evenness); however, it is more informative to study these properties independently [58]. Quantification of alpha-diversity is a crucial step in understanding biodiversity patterns, and these measurements provide insight on how microbial communities (diversity) can be correlated to environmental factors. For example, a global survey of the Pacific Ocean from the Arctic to the Antarctic illustrated the usefulness of these indices and the use of microbial genetic markers in estimating microbial diversity [59]. Lastly, we want to highlight that although it seems intuitive to assume that habitats that have higher diversity would be better (more productive, resilient, and stable), we would like to caution against this perception. There is mounting evidence to indicate that managed landscapes (crop lands and silviculture) are more productive than natural ones.

Beta (β)- and gamma (γ)-diversity

Beta (β)-diversity is a measure of species turnover across samples or a landscape, whereas gamma (γ)-diversity measures the turnover rates across regional/continental scales. It was Whittaker [60] who first defined and formulated how to calculate this turnover across landscapes as total number of taxa at the regional scale (γ) minus the (α) diversity of each sample. This has been broadly understood as the original formula that describes turnover rates, but since then, there have been numerous formulations that have developed. Both measures of diversity have been used widely in macro and microbial ecology; however, there are different indices used in both fields of ecology. Broadly, there are four main classes in which all these indices can be categorized into: (1) Whittaker, (2) Min-Mx, (3) Cody, and (4) Abundance. Whittaker indices are based on the original formula published by Whittaker, and the Min-Max class is a modification version of Whittaker's equations that aims to minimize errors by focusing on the minimum and maximum number of unique taxa between sites. The Cody class of indices focuses on measuring the turnover rates of only unique taxa between sites. Lastly, Abundance-based indices incorporate the abundance values of the taxa in its calculations [61].

To date, there are over a 17 different β-diversity indices that have been described, but all these fall within these four classes mentioned above and have been extensively reviewed in [62]. Despite the plethora of indices that exist, there is mounting concerns among researchers regarding how these indices are quantified. The main concern is that these give a crude estimation of the local richness across sites and their inherent dependency on each other is crucially problematic [57].

Here, we propose a technique widely applied in macro ecological studies, i.e. calculating the total beta-diversity directly from the dataset without quantifying α-diversity dependencies. Briefly, all components of β-diversity are calculated independently and partitioned, i.e. partitioning the dissimilarities that exist as a result of differences in species replacement and partitioning those that results from differences in local species richness at each site/samples [33, 63]. This approach has been shown to be far superior in estimating turnover rates, and in some instances, the ecological questions being asked can be answered by one of the two partitions of β-diversity. There are entanglements of the different α-diversity dependencies that exist when calculating β-diversity using crude indices; moreover, by calculating the independent components, it is possible to correctly discern the causes of the observed diversity patterns. For example, Carvalho et al. [64] highlighted the relative roles of the different β-diversity components (B_{rich} – differences richness; and B_{-3} – difference in species replacement) and how these contribute to the overall biogeographical patterns. Also, Baselga [65] formulated a new approach by partitioning the turnover into two components (B_{ness} [measure of nestedness] and B_{rich}) and he was able to discern the underlying causes of the observed variation in beetle diversity patterns.

Finally, although there are benefits to partitioning biodiversity measurements into its components, it is worth noting that there are some ecological questions that can be answered without partitioning. However, as observed in many microbial ecology studies from extreme environments, surveys are usually conducted to determine the biogeographical patterns across landscapes. Diversity partitioning is especially interesting for researchers who aim to discern how communities differ and to determine which prevailing biodiversity processes are underlying the observed patterns over spatial and temporal scales.

7.3.3 Measuring functional profiles and microbial interactions

The decline in sequencing costs over the past decades has enabled researchers to study how microbial communities vary in their composition and functional profiles across extreme environments (Fig. 7.1). Shotgun metagenomic sequencing has advantages over amplicon-based sequencing in that it circumvents the biases introduced by PCR and, most importantly, offers the possibility to identify gene function as part of the metagenome. In recent years, shotgun metagenomic sequencing has been used in conjunction with amplicon-based approaches [66–68]. These studies have not only highlighted the immense diversity that exists in the environment but also shed light on the potential functional characteristics. Their use is becoming widespread in many microbial studies, but most studies focused on describing communities without looking at interactions among microbial taxa and functional redundancy that may exist within communities. Along with shotgun metagenomic sequencing, another approach has been developed to predict the potential function of microbial communities based on their 16S rRNA gene content using Phylogenetic Investigation of Communities by Reconstruction of Unobserved States (PICRUSt) [69]. This tool works by using 16S rRNA gene sequence data to predict the functional profile of the samples by comparing amplicon sequences with published genomes in databases.

The inherent limitation of this approach is that the predictive power of the algorithm is dependent on the availability of complete genome in the database. To date, the Integrative Human Microbiome Project and its predecessor (Human Microbiome Project) focused on culture-independent methods for the enumeration and characterization of the human microbiome [70, 71]. An unforeseen consequence of these projects was the creation of an extensive database of complete microbial genomes; it is then not surprising that the PICRUSt approach has shown great success in human microbiome studies, but it loses its predictive powers when applied to extreme environments (environments where most microbes have not been described and complete genomes hardly exist) [72, 73]. Furthermore, guild-based approaches have been widely used in extreme environments to characterize community interactions/functional potential, especially for nematode [74], fungi [75], and bacteria [3, 17]. However, such methods often require *priori* knowledge of the populations and information is often inferred from known genomes and derived mainly from culturable components of the community. Similarly, this approach is often criticized for relying on the culturable microbial community as most microbes from extreme environments are yet to be cultured and characterized.

Current models that aim to predict microbial community interactions are steeped in stoichiometric principles, and this offers a limited outlook of microbial diversity and interactions. Where direct evidence of interactions is lacking, the use of proximal data highlights some key pathways and properties [76]. Statistical inferences using Markov networks are emerging as the best predictors of microbial community interactions [77–79]. These models have consistently outperformed other methods in predicting the level and degree of interactions of species in a community from compositional datasets. Briefly, Markov networks and Hidden Markov models work by predicting the possible interactions given the current state; they do this by analyzing all possible/future interactions independent of any *a priori* knowledge. For example, Harris [77] highlighted that when the interaction of two competing taxa are modeled in a network, their negative interactions becomes reversed (positive) with the introduction of a third taxon (one that competes with both). It is this capacity of these models to disentangle all possible interactions in an ecological network that makes them such a superior approach. Similarly, models that use sparse observed data as the background have become widespread in microbial studies; here, inferences are made from observed data to predict the interaction of microbe [80, 81]. These models heavily rely on using statistical decision trees (random forest models) and have been shown to be particularly important when predicting how taxa in a sample would respond given certain conditions (modeling response variable as an integral function of the microbial community dynamics).

Lastly, an approach that reduces the compositional bias created by sequencing datasets and is able to predict correct interactions without being influenced by common factors between microbes would be seen as ideal. A hierarchal Bayesian model that was recently published satisfies this criterion [82]. This approach is able to accurately predict (1) microbe-microbe interactions and (2) microbe-environment interactions from sequence datasets. Although, this model has been applied only to human gut samples, we believe that such a model can be applied to extreme environments to elucidate previously unknown interactions.

7.3.4 Phylogenetic diversity and inference in extreme soils

Microbial phylogenetics provides a deeper understanding of the relationships between taxa and provides bases for inference on trait evolution/organization. In essence, by assessing phylogenies, we can incorporate into our analyses characteristics of closely related taxa even if we cannot observe them directly [83]. The incorporation of phylogenetic analysis in assessing diversity provides insight on how niche shifts can be predicted from phylogenies. To perform any downstream analysis, a phylogenetic tree is required; to this end, there exist separate approaches to their construction. Briefly, homologous sequences are aligned and nucleotide substitution models are used to infer the most likely phylogeny for those sequences. The resultant tree is then used to measure diversity by calculating the total branch length of the tree; samples with higher diversity will have higher sum than less diverse samples [84]. However, most studies that have investigated the phylogenetic diversity of microbes from extreme environments often use the default parameters in their sequence pipelines or make no mention to the parameters used to create the tree [26, 85]. Incorrectly constructed trees tend to reveal incorrect phylogenetic relationship, and any inference from such trees is questionable. It is important to note that certain lineages evolve faster than others, and it is for this reason that a corrected nucleotide substation model must be used [86]. Moreover, the aligned sequences are classified using BLAST-based algorithm to get their identity [87]. BLAST-based pipelines are unable to resolve the taxonomy of sequences efficiently, especially short-read sequences that result from commonly used Illumina sequencing of the 16S rRNA gene. This incorrect taxonomic assignment and incorrect tree construction have huge implication for downstream inferences. For instance, a well-characterized phylogenetic tree (correctly constructed) provides little information regarding the presence of functional genes, and caution must be taken when inference is drawn from prokaryote trees. Moreover, the occurrence of mobile genetic elements and frequency of horizontal gene transfer allow for the evolution/adaptation to new niche spaces. However, this exchange in genetic material often occurs more readily between taxa that are phylogenetically closely related [8, 88]. For example, the detection of the same taxon across a variety of extreme environments does not infer any information about the ecological roles performed by the same taxon in contrasting environments. A new approach that circumvents these issues is phylogenetic placement. Phylogenetic placement algorithms involve the construction of a phylogenetic tree from known genomes in the databank and unto this aligned sample sequences are best placed on node corresponding to the same sequences [89].

This approach allows for the correct taxonomic assignment of microbes from amplicon sequence datasets and allows for the detection of changes in diversity patterns. We believe an approach that reduces uncertainty and biases is best for any phylogenetic inferences. It has recently been applied in human microbiome studies showing better performance than previous methods [90] and recently in soil microbiomes [91].

7.4 Conclusion

Many approaches to describe and quantify microbial diversity are available, enabling us to answer basic questions about biodiversity patterns and ecological processes. However, some approaches are challenging to apply to less-well-known environments because they were devised for environments that are well described. There is a need to improve and adapt these approaches for vulnerable extreme environments, especially in view of the rapid pace of global changes in some of these environments that might change fundamental patterns of microbial diversity.

References

[1] Achtman M, Wagner M. Microbial diversity and the genetic nature of microbial species. Nat Rev Microbiol. 2008;6:431–40. doi: 10.1038/nrmicro1872.
[2] Shade A. Diversity is the question, not the answer. ISME J. 2017;11:1–6. doi: 10.1038/ismej.2016.118.
[3] Torsvik V, Ovreas L. Microbial diversity and function in soil: from genes to ecosystems. Curr Opin Microbiol. 2002;5:240–5. doi: 10.1016/S1369-5274(02)00324-7.
[4] Fierer N, Bradford MA, Jackson RB. Toward an ecological classification of soil bacteria. Ecology. 2007;88:1354–64.
[5] Chen Y, Kuang J, Jia P, et al. Effect of environmental variation on estimating the bacterial species richness. Front Microbiol. 2017;8:690.
[6] Curd EE, Martiny JBH, Li HY, Smith TB. Bacterial diversity is positively correlated with soil heterogeneity. Ecosphere. 2018;9:e02079. doi: 10.1002/ecs2.2079.
[7] Martiny JBH, Bohannan BJM, Brown JH, et al. Microbial biogeography: putting microorganisms on the map. Nat Rev Microbiol. 2006;4:102–12.
[8] Frost LS, Leplae R, Summers AO, Toussaint A. Mobile genetic elements: the agents of open source evolution. Nat Rev Microbiol. 2005;3:722–32. doi: 10.1038/nrmicro1235.
[9] Young IM, Crawford JW, Nunan N, Otten W, Spiers A. Microbial Distribution in Soils: Physics and Scaling. Adv Agron. 2008;100:81–21. doi: 10.1016/S0065-2113(08)00604-4.
[10] Tedersoo L, Bahram M, Polme S, et al. Global diversity and geography of soil fungi. Science. 2014;346:1078–+. doi: Artn 125668810.1126/Science.1256688.
[11] Fierer N, Strickland MS, Liptzin D, Bradford MA, Cleveland CC. Global patterns in belowground communities. Ecol Lett. 2009;12:1238–49. doi: 10.1111/j.1461-0248.2009.01360.x.
[12] Nemergut DR, Schmidt SK, Fukami T, et al. Patterns and processes of microbial community assembly. Microbiol Mol Biol Rev. 2013;77:342–56. doi: 10.1128/mmbr.00051-12.
[13] Dini-Andreote F, Stegen JC, van Elsas JD, Salles JF. Disentangling mechanisms that mediate the balance between stochastic and deterministic processes in microbial succession. P Natl Acad Sci USA. 2015;112:E1326–32. doi: 10.1073/pnas.1414261112.

[14] Tripathi BM, Stegen JC, Kim M, Dong K, Adams JM, Lee YK. Soil pH mediates the balance between stochastic and deterministic assembly of bacteria. ISME J. 2018;12:1072–83. doi: 10.1038/s41396-018-0082-4.
[15] Mäder P, Fliessbach A, Dubois D, Gunst L, Fried P, Niggli U. Soil fertility and biodiversity in organic farming. Science. 2002;296:1694–7.
[16] Schulz S, Brankatschk R, Dümig A, Kögel-Knabner I, Schloter M, Zeyer J. The role of microorganisms at different stages of ecosystem development for soil formation. Biogeosciences. 2013;10:3983–96.
[17] Nannipieri P, Ascher J, Ceccherini M, Landi L, Pietramellara G, Renella G. Microbial diversity and soil functions. Eur J Soil Sci. 2003;54:655–70.
[18] Wardle DA, Bardgett RD, Klironomos JN, Setälä H, Van Der Putten WH, Wall DH. Ecological linkages between aboveground and belowground biota. Science. 2004;304:1629–33.
[19] Moroenyane I, Tripathi B, Dong K, Sherman C, Steinberger Y, Adams J. Bulk soil bacterial community mediated by plant community in Mediterranean ecosystem, Israel. Appl Soil Ecol. 2017;124:104–9.
[20] Neufeld JD, Mohn WW. Unexpectedly high bacterial diversity in arctic tundra relative to boreal forest soils, revealed by serial analysis of ribosomal sequence tags. Appl Environ Microb. 2005;71:5710–8. doi: 10.1128/Aem.71.10.5710-5718.2005.
[21] Gittel A, Barta J, Kohoutova I, et al. Distinct microbial communities associated with buried soils in the Siberian tundra. ISME J. 2014;8:841–53. doi: 10.1038/ismej.2013.219.
[22] WynnWilliams DD. Antarctic microbial diversity: The basis of polar ecosystem processes. Biodivers Conserv. 1996;5:1271–93. doi: 10.1007/Bf00051979.
[23] Smith JJ, Tow LA, Stafford W, Cary C, Cowan DA. Bacterial diversity in three different Antarctic cold desert mineral soils. Microbial Ecol. 2006;51:413–21. doi: 10.1007/s00248-006-9022-3.
[24] Pointing SB, Chan Y, Lacap DC, Lau MCY, Jurgens JA, Farrell RL. Highly specialized microbial diversity in hyper-arid polar desert (vol 106, pg 19964, 2009). Proc Natl Acad Sci USA. 2010;107:1254–1254. doi: 10.1073/pnas.0913882107.
[25] Yergeau E, Newsham KK, Pearce DA, Kowalchuk GA. Patterns of bacterial diversity across a range of Antarctic terrestrial habitats. Environ Microbiol.2007;9:2670–82. doi: 10.1111/j.1462-2920.2007.01379.x.
[26] Chu HY, Fierer N, Lauber CL, Caporaso JG, Knight R, Grogan P. Soil bacterial diversity in the Arctic is not fundamentally different from that found in other biomes. Environ Microbiol. 2010;12:2998–3006. doi: 10.1111/j.1462-2920.2010.02277.x.
[27] Zhalnina K, Dias R, de Quadros PD, et al. Soil pH determines microbial diversity and composition in the park grass experiment. Microbial Ecol. 2015;69:395–406. doi: 10.1007/s00248-014-0530-2.
[28] Hughes JB, Hellmann JJ, Ricketts TH, Bohannan BJ. Counting the uncountable: statistical approaches to estimating microbial diversity. Appl Environ Microb. 2001;67:4399–406.
[29] Roesch LF, Fulthorpe RR, Riva A, et al. Pyrosequencing enumerates and contrasts soil microbial diversity. ISME J. 2007;1: 283–90.
[30] Locey KJ, Lennon JT (2016) Scaling laws predict global microbial diversity. Proceedings of the National Academy of Sciences: 201521291.
[31] Haegeman B, Hamelin J, Moriarty J, Neal P, Dushoff J, Weitz JS. Robust estimation of microbial diversity in theory and in practice. ISME J. 2013;7:1092–101. doi: 10.1038/ismej.2013.10.
[32] Jost L, DeVries P, Walla T, Greeney H, Chao A, Ricotta C. Partitioning diversity for conservation analyses. Divers Distrib. 2010;16:65–76. doi: 10.1111/j.1472-4642.2009.00626.x.
[33] Jost L. Partitioning diversity into independent alpha and beta components (vol 88, pg 2427, 2007). Ecology. 2009;90:3593–3593.
[34] Gilbert JA, Jansson JK, Knight R. The Earth Microbiome project: successes and aspirations. BMC Biol. 2014;12:69. doi: 10.1186/s12915-014-0069-1.

[35] Loman NJ, Misra RV, Dallman TJ, et al. Performance comparison of benchtop high-throughput sequencing platforms. Nat Biotechnol. 2012;30:434–+. doi: 10.1038/nbt.2198.

[36] Scholz MB, Lo CC, Chain PSG. Next generation sequencing and bioinformatic bottlenecks: the current state of metagenomic data analysis. Curr Opin Biotech. 2012;23:9–15. doi: 10.1016/j.copbio.2011.11.013.

[37] Tripathi R, Sharma P, Chakraborty P, Varadwaj PK. Next-generation sequencing revolution through big data analytics. Front Life Sci. 2016;9:119–49. doi: 10.1080/21553769.2016.1178180.

[38] Gloor GB, Reid G. Compositional analysis: a valid approach to analyze microbiome high-throughput sequencing data. Can J Microbiol. 2016;62:692–703. doi: 10.1139/cjm-2015-0821.

[39] Walker AW, Martin JC, Scott P, Parkhill J, Flint HJ, Scott KP. 16S rRNA gene-based profiling of the human infant gut microbiota is strongly influenced by sample processing and PCR primer choice. Microbiome. 2015;3:26. doi: 10.1186/s40168-015-0087-4.

[40] Alatalo RV. Problems in the measurement of evenness in ecology. Oikos. 1981;37:199–204. doi: 10.2307/3544465.

[41] Hurlbert SH. The nonconcept of species diversity: a critique and alternative parameters. Ecology. 1971;52:577–86.

[42] MacArthur R. On the relative abundance of species. Am Nat. 1960;94:25–36.

[43] Zhang ZJ, Qu YY, Li SZ, et al. Soil bacterial quantification approaches coupling with relative abundances reflecting the changes of taxa. Sci Rep-Uk. 2017;7:4837. doi: 10.1038/s41598-017-05260-w.

[44] Props R, Kerckhof FM, Rubbens P, et al. Absolute quantification of microbial taxon abundances. ISME J. 2017;11:584–7. doi: 10.1038/ismej.2016.117.

[45] Shi PX, Zhang AR, Li HZ. Regression analysis for microbiome compositional data. Ann Appl Stat. 2016;10:1019–40. doi: 10.1214/16-Aoas928.

[46] Morton JT, Sanders J, Quinn RA, et al. Balance trees reveal microbial niche differentiation. Msystems. 2017;2:e00162–16. doi: 10.1128/mSystems.00162-16.

[47] Li ZG, Lee K, Karagas MR, et al. Conditional regression based on a multivariate zero-inflated logistic-normal model for microbiome relative abundance data. Stat Biosci. 2018;10:587–608. doi: 10.1007/s12561-018-9219-2.

[48] Smets W, Leff JW, Bradford MA, McCulley RL, Lebeer S, Fierer N. A method for simultaneous measurement of soil bacterial abundances and community composition via 16S rRNA gene sequencing. Soil Biol Biochem. 2016;96:145–51. doi: 10.1016/j.soilbio.2016.02.003.

[49] Pace NR. A molecular view of microbial diversity and the biosphere. Science. 1997;276:734–40. doi: 10.1126/science.276.5313.734.

[50] Borneman J, Triplett EW. Molecular microbial diversity in soils from eastern Amazonia: evidence for unusual microorganisms and microbial population shifts associated with deforestation. Appl Environ Microb. 1997;63:2647–53.

[51] DeSantis TZ, Hugenholtz P, Larsen N, et al. Greengenes, a chimera-checked 16S rRNA gene database and workbench compatible with ARB. Appl Environ Microbiol. 2006;72:5069–72.

[52] Chun J, Lee JH, Jung Y, et al. EzTaxon: a web-based tool for the identification of prokaryotes based on 16S ribosomal RNA gene sequences. Int J Syst Evol Micr. 2007;57: 2259–61. doi: 10.1099/ijs.0.64915-0.

[53] Yoon SH, Ha SM, Kwon S, Lim J, Kim Y, Seo H, Chun J. Introducing EzBioCloud: a taxonomically united database of 16S rRNA gene sequences and whole-genome assemblies. Int J Syst Evol Micr. 2017;67:1613–7. doi: 10.1099/ijsem.0.001755.

[54] Case RJ, Boucher Y, Dahllof I, Holmstrom C, Doolittle WF, Kjelleberg S. Use of 16S rRNA and rpoB genes as molecular markers for microbial ecology studies. Appl Environ Microb. 2007;73:278–88. doi: 10.1128/Aem.01177-06.

[55] Galimberti A, Spada M, Russo D, et al. Integrated operational taxonomic units (IOTUs) in echolocating bats: a bridge between molecular and traditional taxonomy. PloS One. 2012;7:e40122. doi: 10.1371/journal.pone.0040122.
[56] Callahan BJ, McMurdie PJ, Holmes SP. Exact sequence variants should replace operational taxonomic units in marker-gene data analysis. ISME J. 2017;11:2639–43. doi: 10.1038/ismej.2017.119.
[57] Jost L. Entropy and diversity. Oikos. 2006;113:363–75. doi: 10.1111/j.2006.0030-1299.14714.x.
[58] Jost L. Independence of alpha and beta diversities. Ecology. 2010;91:1969–U1104. doi: 10.1890/09-0368.1.
[59] Baldwin AJ, Moss JA, Pakulski JD, Catala P, Joux F, Jeffrey WH. Microbial diversity in a Pacific Ocean transect from the Arctic to Antarctic circles. Aquat Microb Ecol. 2005;41:91–102. doi: 10.3354/ame041091.
[60] Whittaker RH. Vegetation of the Siskiyou mountains, Oregon and California. Ecol Monogr. 1960;30:279-38.
[61] Schroeder PJ, Jenkins DG. How robust are popular beta diversity indices to sampling error? Ecosphere. 2018;9:e02100. doi: 10.1002/ecs2.2100.
[62] Tuomisto H. A diversity of beta diversities: straightening up a concept gone awry. Part 2. Quantifying beta diversity and related phenomena. Ecography. 2010;33:23–45. doi: 10.1111/j.1600-0587.2009.06148.x.
[63] Jost L. Partitioning diversity into independent alpha and beta components. Ecology. 2007;88:2427–39. doi: 10.1890/06-1736.1.
[64] Carvalho JC, Cardoso P, Gomes P. Determining the relative roles of species replacement and species richness differences in generating beta-diversity patterns. Global Ecol Biogeogr. 2012;21:760–71. doi: 10.1111/j.1466-8238.2011.00694.x.
[65] Baselga A. Partitioning the turnover and nestedness components of beta diversity. Global Ecol Biogeogr. 2010;19:134–43. doi: 10.1111/j.1466-8238.2009.00490.x.
[66] Ong SH, Kukkillaya VU, Wilm A, et al. Species identification and profiling of complex microbial communities using shotgun illumina sequencing of 16S rRNA amplicon sequences. Plos One. 2013;8:e60811. doi: 10.1371/journal.pone.0060811.
[67] Fierer N, Leff JW, Adams BJ, et al. Cross-biome metagenomic analyses of soil microbial communities and their functional attributes. Proc Natl Acad Sci USA. 2012;109:21390-5. doi: 10.1073/pnas.1215210110.
[68] Bulgarelli D, Garrido-Oter R, Munch PC, et al. Structure and function of the bacterial root microbiota in wild and domesticated barley. Cell Host Microbe. 2015;17:392–403. doi: 10.1016/j.chom.2015.01.011.
[69] Langille MGI, Zaneveld J, Caporaso JG, et al. Predictive functional profiling of microbial communities using 16S rRNA marker gene sequences. Nat Biotechnol. 2013;31:814–+. doi: 10.1038/nbt.2676.
[70] Aagaard K, Petrosino J, Keitel W, et al. The Human Microbiome Project strategy for comprehensive sampling of the human microbiome and why it matters. Faseb J. 2013;27:1012–22. doi: 10.1096/fj.12-220806.
[71] Peterson J, Garges S, Giovanni M, et al. The NIH Human Microbiome Project. Genome Res. 2009;19:2317–23. doi: 10.1101/gr.096651.109.
[72] Goodrich JK, Waters JL, Poole AC, et al. Human genetics shape the gut microbiome. Cell. 2014;159:789–99. doi: 10.1016/j.cell.2014.09.053.
[73] Iwai S, Weinmaier T, Schmidt BL, et al. Piphillin: improved prediction of metagenomic content by direct inference from human microbiomes. PloS One. 2016;11:e0166104. doi: 10.1371/journal.pone.0166104.
[74] Kerfahi D, Park J, Tripathi BM, et al. Molecular methods reveal controls on nematode community structure and unexpectedly high nematode diversity, in Svalbard high Arctic tundra. Polar Biol. 2017;40:765–76. doi: 10.1007/s00300-016-1999-6.

[75] Di Lonardo DP, Pinzari F, Lunghini D, Maggi O, Granito VM, Persiani AM. Metabolic profiling reveals a functional succession of active fungi during the decay of Mediterranean plant litter. Soil Biol Biochem. 2013;60:210–9. doi: 10.1016/j.soilbio.2013.02.001.

[76] Widder S, Allen RJ, Pfeiffer T, et al. Challenges in microbial ecology: building predictive understanding of community function and dynamics. ISME J. 2016;10:2557–68. doi: 10.1038/ismej.2016.45.

[77] Harris DJ. Inferring species interactions from co-occurrence data with Markov networks. Ecology. 2016;97:3308–14. doi: 10.1002/ecy.1605.

[78] Love WJ, Zawack KA, Booth JG, Grohn YT, Lanzas C. Markov networks of collateral resistance: National Antimicrobial Resistance Monitoring System Surveillance results from *Escherichia coli* isolates, 2004–2012. Plos Comput Biol. 2016;12:e1005160. doi: 10.1371/journal.pcbi.1005160.

[79] Fernandez M, Riveros JD, Campos M, Mathee K, Narasimhan G. Microbial "social networks". BMC Genom. 2015;16:S6. doi: 10.1186/1471-2164-16-S11-S6.

[80] Edwards JA, Santos-Medellin CM, Liechty ZS, et al. Compositional shifts in root-associated bacterial and archaeal microbiota track the plant life cycle in field-grown rice. PloS Biol. 2018;16:e2003862. doi: 10.1371/journal.pbio.2003862.

[81] Kurtz ZD, Muller CL, Miraldi ER, Littman DR, Blaser MJ, Bonneau RA. Sparse and compositionally robust inference of microbial ecological networks. Plos Comput Biol. 2015;11:e1004226. doi: 10.1371/journal.pcbi.1004226.

[82] Yang YQ, Chen N, Chen T. Inference of environmental factor-microbe and microbe-microbe associations from metagenomic data using a hierarchical Bayesian statistical model. Cell Syst. 2017;4:129–+. doi: 10.1016/j.cels.2016.12.012.

[83] Washburne AD, Morton JT, Sanders J, et al. Methods for phylogenetic analysis of microbiome data. Nat Microbiol. 2018;3:652–61. doi: 10.1038/s41564-018-0156-0.

[84] Faith DP. Conservation evaluation and phylogenetic diversity. Biol Conserv. 1992;61:1–10. doi: 10.1016/0006-3207(92)91201-3.

[85] Junge K, Imhoff F, Staley T, Deming JW. Phylogenetic diversity of numerically important arctic sea-ice bacteria cultured at subzero temperature. Microb Ecol. 2002;43:315–28. doi: 10.1007/s00248-001-1026-4.

[86] Lemmon AR, Moriarty EC. The importance of proper model assumption in Bayesian phylogenetics. Syst Biol. 2004;53:265–77. doi: 10.1080/10635150490423520.

[87] McGinnis S, Madden TL. BLAST: at the core of a powerful and diverse set of sequence analysis tools. Nucleic Acids Res. 2004;32:W20–5. doi: 10.1093/nar/gkh435.

[88] Soucy SM, Huang JL, Gogarten JP. Horizontal gene transfer: building the web of life. Nat Rev Genet. 2015;16:472–82. doi: 10.1038/nrg3962.

[89] Berger SA, Krompass D, Stamatakis A. Performance, accuracy, and web server for evolutionary placement of short sequence reads under maximum likelihood. Syst Biol. 2011;60:291–302. doi: 10.1093/sysbio/syr010.

[90] Srinivasan S, Hoffman NG, Morgan MT, et al. Bacterial communities in women with bacterial vaginosis: high resolution phylogenetic analyses reveal relationships of microbiota to clinical criteria. PloS One. 2012;7:e37818. doi: 10.1371/journal.pone.0037818.

[91] Mahe F, de Vargas C, Bass D, et al. Parasites dominate hyperdiverse soil protist communities in Neotropical rainforests. Nat Ecol Evol. 2017;1:0091. doi: 10.1038/s41559-017-00911.

Index

https://doi.org/10.1515/9783110525786-008

Printed and bound by CPI Group (UK) Ltd, Croydon, CR0 4YY

13/09/2026

14960851-0008